sto

7-2-93

TAXONOMIC ATLAS
of
LIVING PRIMATES

TARSIUS SYRICHTA

TAXONOMIC ATLAS

of

LIVING PRIMATES

by

A. B. CHIARELLI

Institute of Anthropology
Primatology Centre
University of Turin, Italy

1972

ACADEMIC PRESS
LONDON AND NEW YORK

ACADEMIC PRESS INC. (LONDON) LTD
24/28 Oval Road
London NW1

U.S. Edition published by
ACADEMIC PRESS INC.
111 Fifth Avenue,
New York, New York 10003

Library of Congress Catalog Card Number: 79–129785
ISBN: 0–12–172550–2

PRINTED IN GREAT BRITAIN BY
W & J Mackay Limited, Chatham

PREFACE

Owing to the ever-increasing use of non-human Primates in experimental investigations, it is essential to have easy accessibility to up-to-date records of the biological data. This volume consists of a list of species with their synonyms and non-Latin names, and further atlases are contemplated.

Consistency in nomenclature of the Primates is extremely important for Zoological Gardens: the use of different names or synonyms causes confusion to the visitors and creates difficulties of communication with organizations involved in animal exchange.

Several excellent textbooks are available to students and Zoo directors for reference, including the monumental treatise by Hill, von Fiedler's chapter in *Primatologia*, and the valuable book by Napier and Napier.

The present volume is not intended as an alternative to the above treatises, and in fact much information was derived from them. It is intended as a simple guide that can serve as the taxonomic basis for the set of atlases envisaged, which will provide biological data on the different Primate species.

In the classification key for the various species only the readily recognizable external features have been taken into consideration. The key is organized into a system of four decimals: the first refers to the sub-order, the second to the family, the third to the genus, and the fourth to the species. Subspecies have not been taken into consideration, and the subspecies synonyms have therefore almost always been omitted. It has been attempted to give complete lists of synonyms in order to provide access to early literature.

Non-Latin names given here are in English, German, French, Spanish and Italian. Only occasionally have local names (i.e. those given by the natives) been reported.

I am already prepared for the criticism that such a work will have for its over simplification of taxonomy and for the unavoidable mistakes it certainly contains, especially in the list of synonyms.

In the preparation of this atlas, I am much indebted to Miss Maria Monchietto for her efficient help in the preparation of the distribution maps and for the correction of the proofs; to many eminent primatologists from whom I obtained information and suggestions for preparing the taxonomic key and the lists of synonyms and vernacular names; to all friends, Agencies, Primatological Centres and Zoological Gardens which helpfully provided me with the illustrations; and last but not least to the Production Department of Academic Press, and particularly Mrs. Barbara Renvoize, for their cooperation.

December, 1971

A. B. Chiarelli

REFERENCES TO MAIN SOURCES OF INFORMATION

Fiedler, W. von (1956). Übersicht über das system der Primates. *In* "Primatologia", Vol I (Eds. H. Hofer, A. M. Schultz and D. Starck). S. Karger, Basle and New York.

Hershkovitz, P. (1966). Taxonomic notes on Tamarins, genus Sanguinus (Callithricidae, Primates), with descriptions of four forms. *Folia primat.* **4**, 381–395. S. Karger, Basel and New York.

Hershkovitz, P. (1968). Metachronism or the principle of evolutionary change in mammalian tegumentary colors. *Evolution* **22**, 556–575.

Hill, W. C. Osman (1953). "Primates—Comparative Anatomy and Taxonomy. Vol I Strepsirrhini". Edinburgh University Press.

Hill, W. C. Osman (1955). "Primates—Comparative Anatomy and Taxonomy. Vol II Haplorhini: Tarsioidea". Edinburgh University Press.

Hill, W. C. Osman (1957). "Primates—Comparative Anatomy and Taxonomy. Vol III Pithecoidea Platyrrhini". Edinburgh University Press.

Hill, W. C. Osman (1960). "Primates—Comparative Anatomy and Taxonomy. Vol IV Cebidae. Part A". Edinburgh University Press.

Hill, W. C. Osman (1962). "Primates—Comparative Anatomy and Taxonomy. Vol V. Cebidae. Part B". Edinburgh University Press.

Hill, W. C. Osman (1966). "Primates—Comparative Anatomy and Taxonomy. Vol VI. Catarrhini. Cercopithecoidea". Edinburgh University Press.

Hill, W. C. Osman (1967). The Genera of Old World Apes and Monkeys. *In* Taxonomy and Phylogeny of Old World Primates with references to the origin of Man. Rosenberg e Sellier, Turin.

Kingdon, J. (1971). "East African Mammals", Vol. 1. Academic Press, London and New York.

Napier, J. R. and Napier, P. H. (1967). "A Handbook of Living Primates". Academic Press, London and New York.

CONTENTS

1.(1–7)
    (A) With a rhinarium (i.e. a wet nose-pad like
         that of dog) .... 1.(1–6)
    (B) Without a rhinarium, nose dry .... 1.7. TARSIIDAE

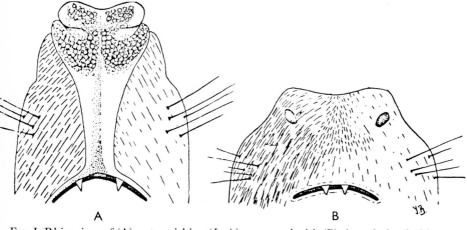

         A   B

FIG. I. Rhinarium of (A) a strepsirhine (*Loris*) compared with (B) that of a haplorhine (*Tarsius*). From Hill (1953). "Primates—Comparative Anatomy and Taxonomy", Vol. I, p. 39.

1.(1–6)
    (A) Squirrel-like animals with small bodies
         and short arms and legs; tail bushy or
         tufted (*Ptilocercus*) and approximately
         equal to or sometimes longer than head
         and body length; ears small and bare
         and set close to head; short whiskers;
         elongated shrew-like nose terminating in
         a naked, moist snout. Geographically
         distributed throughout South East Asia* 1.1. TUPAIIDAE
    (B) Ear with double supratragus and a
         mechanism for permitting its scapha to be
         folded; no vibrissae on fore limb; re-
         duced facial vibrissae; globular skull

* Some authors (Hill, Martin, etc.) do not include this group among the Primates; the group nevertheless represent important evolutionary developments towards the Primates.

with shortened face; large orbits; more than one pair of mammary glands. Geographically distributed in South East Asia and Africa, but not in Madagascar 1.(2–3)

(C) Ear with a single supratragus, scapha incapable of folding; a full complement of facial vibrissae and in many cases carpal vibrissae in addition; special cutaneous gland often present on forelimb and in the anal region; skull elongated with brain case usually dorso-ventrally compressed; orbits small to medium (except Cheirogaleinae); mammary glands usually reduced to a single pectoral pair (except Cheirogaleinae and Daubentonidae). Geographically distributed in Madagascar and the adjacent Islands........... 1.(4–6)

1.(2–3)
(A) Limbs subequal in length, ears smaller, partly or wholly hidden by the fur; locomotion deliberately slow, climbing by hand-over-hand action; unable to leap .... 1.2. LORISIDAE

(B) Pelvic limbs much longer than pectoral; ears very large; locomotion rapid and effected by saltation................. 1.3. GALAGIDAE

1.(4–6)
(A) Size small to medium; interramal and carpel vibrissae present; tail long with a general tendency to bushiness; ears short or of only moderate length; not naked and generally hidden in the fur .... 1.4. LEMURIDAE

(B) Very large size; interramal and carpal vibrissae absent. Very specialized dentition, upper incisors large............ 1.5. INDRIIDAE

(C) Size medium; interramal and carpal vibrissae absent. Great elongation of manual digits, third one thinner. Large diastema between incisors and molars; mammary glands in a single inguinal pair.............................. 1.6. DAUBENTONIIDAE

2

1.1. (1–4) TUPAIIDAE

(A) Tail naked at the base, but with a horizontally compressed fringe of long hair near the tip, giving tail appearance of a feather. Ears large and membranaceous. Foot pads relatively large and soft...... 1.1.4.1. *Ptilocercus lowii*

(B) Tail hairy to the base throughout its entire extent, usually bushy. Ears small and cartilaginous. Foot pads of moderate development.

    (a) tail hair short, 2–4 mm, only in the centre and lying flat so that the tail appears thin and rat-like. Small species (head and body length 120 mm) .... 1.1.2.(1–2) *Dendrogale*

    (b) tail hair long, at least 10 mm, so that the tail is bushy. Species of various sizes........................ 1.1.1.(1–9) *Tupaia*

    (c) tail hair long, rather close, dark brown fur with yellowish or rufous under parts (very similar to *Tupaia montana*); large body size (largest tupaiid) 1.1.3.1. *Urogale everetti*

1.1.2.(1–2) *Dendrogale*

(A) Grey–brown coat with lighter underparts: face with a conspicuous dark streak running from nose to ears with a lighter stripe above and below.............. 1.1.2.1. *D. murina*

(B) Dark grey coat with an orange mark on the cheek and around the eye.......... 1.1.2.2. *D. melanura*

1.1.1.(1–11) *Tupaia*

(A) With a black stripe down the middle of the back, at least on the front part of the body:

    (a) black stripe continues to the root of the tail; rather small species, hind foot about 40 mm.............. 1.1.1.8. *T. dorsalis*

    (b) black stripe continues for only about half the length of the back and is lost in the dark fur of the posterior half;

larger species; hind foot 45–50 mm . . 1.1.1.7. *T. tana*

(B) Without black stripe, although the fur of the dorsal crest of hair may be somewhat darker than that of sides:

   (a) larger species, head and body length over 140 mm, tail not noticeably longer than head plus body, and usually somewhat shorter:

      i. very large, hind foot about 50 mm; shoulder stripe and stomach bright buff or reddish–brown in contrast to dark brown fur of back; northern forms olive with white underparts . . . . 1.1.1.1. *T. glis*

      ii. smaller, hind foot about 40 mm; shoulder stripe and stomach not much paler than dull brown dorsum . . . . 1.1.1.2. *T. montana*

   (b) smaller species, head and body length under 140 mm, tail at least 25% longer than length of body and head:

      i. hind foot long, about 38–40 mm . . 1.1.1.6. *T. gracilis*

      ii. hind foot short, about 28–32 mm. 1.1.1.5. *T. minor*

      iii. tail distinctly longer than head and body length

         (*a*) living in Nicobar . . . . 1.1.1.3. *T. nicobarica*
                                      and

         (*b*) living in Java . . . . 1.1.1.4. *T. javanica*

(C) Fur red–brown or grey–brown coarsely speckled with black; large hairy ears. . . . 1.1.1 9. *T. ellioti*

1.2.(1–4) Lorisidae

(A) Small size; muzzle more pointed; body slender:

   (a) limbs elongated, excessively slender; index finger present . . . . 1.2.1.1. *Loris tardigradus*

   (b) limbs shorter and stouter, index finger absent . . . . 1.2.3.1. *Arctocebus calabarensis*

(B) Large size; muzzle blunt; body stout:
    (a) tail present; index finger reduced to a stump; spinous processes of cervical vertebrae project above skin level ... *1.2.4.1. Perodicticus potto*

    (b) tail absent or reduced to a stump; index finger present; no projection of vertebrae project above skin level... *1.2.4.1. Perodicticus*

1.2.2.(1–2) *Nycticebus*

(A) Coat colour brown or reddish–brown or grey with white line between eyes from forehead to muzzle, dark marking around eyes and a dark medial stripe extending on the crown where it may diverge to ears and join the dark eye marking .... *1.2.2.1. N. coucang*

(B) Coat colour uniformly brown or reddish–brown or grey. Body size about half that of *N. coucang* .... *1.2.2.2. N. pygmaeus*

1.3.(1) GALAGIDAE

Only one genus .... *1.3.1.(1–5) Galago*

1.3.1.(1–5) *Galago*

(A) Adult body weight over 1000g; head and body length over 300 mm; tail length around 450 mm; muzzle long and robust; tail thick and bushy .... *1.3.1.2. G. crassicaudatus*

(B) Adult body weight not over 300 g; head and body length around 150 mm; tail length around 230 mm; muzzle short:
    (a) underparts yellowish–white or greyish rusty tinge on outside of limbs; tail thick and bushy .... *1.3.1.3. G. alleni*

    (b) no lighter under parts of the body; no rusty tinge of limbs; a distinct light-coloured inter-ocular stripe extending down nose; tail thinly haired; nails specialized, but with the keel

and the sharp point not so strongly
marked......................... 1.3.1.1. *G. senegalensis*
(c) dense and woolly fur of reddish–
brown to pale cinnamon; dark dorsal
median stripe; under parts pale grey;
nails specialized and bearing a central keel ending in a sharp point;
general colour cinnamon; tail thick 1.3.1.5. *G. elegantulus*
(C) Adult body smaller than in the other
species (total length of the head, body
and tail not more than 350 mm); larger
eyes forward facing with pale streak be-
tween them, running down the nose .... 1.3.1.4. *G. demidovii*

1.4.(1–6) LEMURIDAE

(A) Size from small to medium; lower lateral
incisors elongated; head roundish with
short muzzle; tail longer than head and
body. Eyes very large .... 1.4.(1–3) Cheirogaleinae
(B) Size from medium to large (the size of a
cat); head elongated and bushy with fox-
like face (except *Hapalemur*); eyes of
normal dimension .... 1.4.(4–6) Lemurinae

1.4.(1–3) Cheirogaleinae

(A) Size small (head and body 130–250 mm);
ears long, membranaceous; foremost up-
per premolar shorter than the next ..... 1.4.1.(1–2) *Microcebus*
(B) Size larger (total length, head and body,
300–600 mm); ears short, hidden in fur;
foremost upper premolar longer than the
next:
(a) foremost upper premolar caniniform. 1.4.3.1. *Phaner furcifer*
(b) foremost upper premolar not defin-
itely caniniform................. 1.4.2.(1–2) *Cheirogaleus*

1.4.1.(1–2) *Microcebus*

(A) Face marked by a white median nasal

6

stripe. Head and body length 130 mm;
tail length 140 mm .... 1.4.1.1. *M. murinus*
(B) Median nasal stripe absent; head and
body length 250 mm; tail length 280 mm 1.4.1.2. *M. coquerelli*

1.4.2.(1–3) *Cheirogaleus*
(A) Large size......................... 1.4.2.1. *C. major*
(B) Smaller size; tail thicker than in *C. major*
underpart coat colour completely white . 1.4.2.2. *C. medius*

1.4.(4–6) Lemurinae
(A) Tail as long as, or longer than, head and
body; permanent upper incisors present:
(a) premaxilla shortened; mammary
glands in two pairs .... 1.4.4.(1–2) *Hapalemur*
(b) premaxilla not shortened; a single
pair of mammary glands.......... 1.4.5.(1–5) *Lemur*
(B) Tail shorter than head and body, per-
manent upper incisors lacking .... 1.4.6.1. *Lepilemur*
                                                                      mustelinus

1.4.4.(1–2) *Hapalemur*
(A) Size medium (700 mm long); diastema
between canine and premolar; ante-
brachial gland present............... 1.4.4.1. *H. griseus*
(B) Size large (900 mm long) no diastema
behind canine; antebrachial gland absent 1.4.4.2. *H. simus*

1.4.5.(1–5) *Lemur*
(A) Carpal gland and horny spur present on
forearm; scrotum naked; tail ringed alter-
nately black and white .... 1.4.5.1. *L. catta*
(B) No brachial or carpal glandular areas;
scrotum hairy; tail not ringed in adult:
(a) pelage variegated black and white or
black, red and white; head with ruff;
tail shorter than or about the same
length as body................... 1.4.5.2. *L. variegatus*
(b) head without ruff; colour uniform;
tail longer than body:
i. cheek-fringe and ear-tufts present;
sexes differing in pelage: ♂ black,
♀ brown or reddish .... 1.4.5.3. *L. macaco*

7

ii. no cheek-fringe or ear-tufts; sexes
alike in colour:
    (a) ears concealed in fur, hairy
on both sides; general colour
red–brown with limbs rufous 1.4.5.4. *L. rubriventer*
    (b) ears not wholly concealed;
hairy on cranial side, naked
within; sides of muzzle whit-
ish or generally black ...... 1.4.5.5. *L. mongoz*

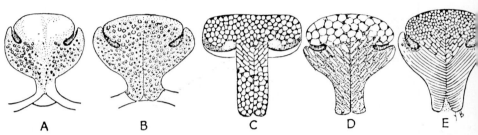

FIG. II. Series of strepsirhine rhinaria, showing evolution of the surface pattern. A, *Periodicticus*; B, *Nycticebus*; C. *Lemur*; D, *Phaner*; E, *Microcebus*. From Osman Hill (1953). "Primates-Comparative Anatomy and Taxonomy", Vol. I, p. 42.

1.5.(1–3) INDRIIDAE

(A) Tail long; size moderate to small:
    (a) size moderate; pelage largely white,
more silky than woolly .... 1.5.1.(1–2) *Propithecus*
    (b) size small; pelage uniformly brown,
or greyish, with white stripes, woolly 1.5.2.1. *Avahi laniger*
(B) Tail a mere stump; size large .... 1.5.3.1. *Indri indri*

1.5.1.(1–2) *Propithecus*

(A) Coat colour dark with strongly melanistic
forms; tail not usually projecting beyond
heel .... 1.5.1.1. *P. diadema*
(B) Coat colour pale–grey or white; body
size smaller than *P. diadema*; tail project-
ing beyond heel .... 1.5.1.2 *P. verreauxi*

1.6.(1) DAUBENTONIIDAE

Only one genus and one species .... 1.6.1.1. *Daubentonia madagascariensis*

1.7.(1) Tarsiidae

Only one genus with three species......... 1.7.1.(1–3) *Tarsius*

1.7.1.(1–3) *Tarsius*

(A) Tail with minute bristles beneath, ar-
ranged in groups of three outlining a
scale-like pattern; a white postauricular
spot; tail tuft, thicker and darker in
respect to the other species, extends about
half way along the tail; dark grey fur
mottled with brown; tarsi hairy........ 1.7.1.1. *T. spectrum*

(B) Tail naked and smooth beneath; no
white post-auricular spots; tarsi virtually
naked:

   (a) tail   macroscopically   completely
smooth beneath; grey fur tinged with
red–brown; tuft of tail with very fine,
soft, short hairs .... 1.7.1.2. *T. syrichta*

   (b) tail with papillary ridges beneath,
grey fur mottled with golden brown;
tuft of the tail with strong brown
hairs .... 1.7.1.3. *T. bancanus*

SIMIAE PLATYRRHINAE: family key

Monkeys with large oral nares directed laterally and (except in *Aotes* and *Brachyteles*) separated by a broad internarial septum; with thumb imperfectly opposable; lacking cheek pouches and ischiatic callosities.
- (A) Digits with claw-like nails except on hallux 2.(1–2)
- (B) Digits with typical nails . . . .   2.3. CEBIDAE

2.(1–2)

- (A) Dental formula $\dfrac{\text{I2}}{2}, \dfrac{\text{C1}}{1}, \dfrac{\text{P3}}{3}, \dfrac{\text{M2}}{2} = 32$ . . . 2.1. CALLITHRICIDAE

- (B) Dental formula $\dfrac{\text{I2}}{2}, \dfrac{\text{C1}}{1}, \dfrac{\text{P3}}{3}, \dfrac{\text{M3}}{3} =: 36$ . . . 2.2. CALLIMICONIDAE

2.1.(1–4) CALLITHRICIDAE

- (A) With elongated lower incisors and incisiform lower canines . . . . 2.1.(1–2)
- (B) With lower canine–incisor relationship normal . . . . 2.1.(3–4)

2.1.(1–2)
- (A) Size normal; mandible high in proportion to its length; ears naked or with long tufts. . . . . 2.1.1.(1–3) *Callithrix*
- (B) Size small; ears untufted, hidden in hair at side of head . . . . 2.1.2.1. *Cebuella pygmaea*

2.1.(3–4)
- (A) Hands with elongated narrow palm and digits III and IV partly syndactylous; pelage forming a mane concealing the ears . . . . 2.1.3.(1–3) *Leontideus*
- (B) Hands with comparatively broad palm, with short digits not connected by interdigital webs, or only extremely narrow ones. Side and crown of head completely covered by hair or nearly bare; ears partly or wholly exposed. . . . . 2.1.4.(1–14) *Saguinus*

10

2.1.1.(1–3) *Callithrix*
  (A) Body fur silvery-white or light brown;
      ears and face bare, crimson or pink; tail
      black, not ringed .... 2.1.1.1. *C. argentata*
  (B) Body fur brownish–black, grizzled; ears
      white or with buff tufts growing from
      pinna; face pigmented with a pink muz-
      zle; tail black at base, faintly ringed
      distally......................... 2.1.1.2. *C. humeralifer*
  (C) Body fur black and grey or brown marbl-
      ing; ears with white, yellow or black
      tufts, face hairy; tail ringed, black and
      grey............................. 2.1.1.3. *C. jacchus*

2.1.3.(1–3) *Leontideus*
  (A) Coat reddish gold all over .... 2.1.3.1. *L. rosalia*
  (B) Coat generally black with golden mane,
      arms and upper tail .... 2.1.3.2. *L. chrysomelas*
  (C) Coat wholly black except for gold fore-
      head and inner side of hind limbs....... 2.1.3.3. *L. chrysopygus*

2.1.4.(1–14) *Saguinus*
  (A) Forehead and crown naked or sparsely
      clothed with short hair to level of ears;
      ears large, exposed, with lamina of
      postero-inferior margin complete and
      rounded .... 2.1.4.(1–8)
  (B) Forehead and crown with conspicuous
      band or crest of long, white hair; rest of
      face and crown bald; ears small, exposed,
      with lamina of postero-inferior margin
      deeply emarginate or obsolete .... 2.1.4.(9–10)
  (C) Sides of head and crown to cheeks and
      chin covered with dark (brown, black or
      reddish) hair; ears large, conspicuous ... 2.1.4.(11–14)

2.1.4.(1–8) *Saguinus*
  (A) Moustache absent:
      (a) body fur black with rust coloured
          marbling on the back .... 2.1.4.1. *S. tamarin*
      (b) as above but with orange or yellow
          hands and feet .... 2.1.4.2. *S. midas*

(B) Conspicuous white moustache:
    (a) face entirely blackish, the muzzle (ex-
    cept the nose) being covered with
    short white hair:
       i. body dominantly whitish or yel-
       lowish–white, showing on the
       dorsum a three-zoned pattern.
       Saddle on middle and lower back
       marbled, striated or vermiculate;
       naked parts of face, ears, hands
       and feet black................ 2.1.4.3. *S. fuscicollis*
      ii. body colour various, never en-
      tirely whitish (or blackish), with
      a two-zonal pattern on the dorsum:
        (*a*) mantle agouti or melano-
        agouti and truncate or tapered
        behind; lower back, rump,
        thighs and underparts domin-
        antly olivaceous or buff–
        brown .... 2.1.4.4. *S. graellsi*
        (*b*) mantle uniformly black and
        tapered behind; lower back,
        rump and thighs dominantly
        reddish or mahogany, under-
        part reddish more or less
        mixed with black .... 2.1.4.5. *S. nigricollis*
    (b) face blackish with an unpigmented
    area covered with longer white hair
    on the muzzle generally including the
    nostrils:
       i. grey with reddish tail and long
       dropping white moustache...... 2.1.4.6. *S. imperator*
      ii. mainly black with conspicuous
      white moustache in the shape of a
      clover-leaf .... 2.1.4.7. *S. mystax*
     iii. a narrow white area around lips,
      blackish–brown back, black limbs
      and tail; underparts and inner
      sides of limbs orange–red .... 2.1.4.8. *S. labiatus*

2.1.4.(9–10) *Saguinus* (*Oedipomidas*)
    (A) Back marbled with buff, crest of white
    hair short and reddish on back of head.. 2.1.4.9. *S. geoffroyi*

(B) Back dark brown with white underparts and limbs; crest of white hair long and flowing .... 2 1.4.10. *S. oedipus*

2.1.4.(11–14) *Saguinus (Marikina)*
  (A) Face with long silvery hair on cheeks, fore-arms and hands whitish........... 2.1.4.11. *S. leucopus*
  (B) Face bare and black as far as vertex:
    (a) coat colour mainly black with brown on back and flanks and some unpigmented areas on the bare face ...... 2.1.4.12. *S. inustus*
    (b) coat colour dichromatic with sharply defined white chest and forelimbs in contrast to yellowish–brown hind parts .... 2.1.4.13. *S. bicolor*
    (c) coat colour brown with fore-arms and hands yellowish .... 2.1.4.14. *S. martinsi*

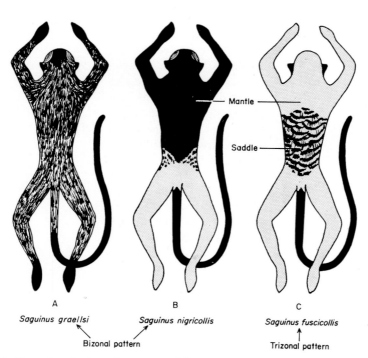

FIG. III. Bizonal and trizonal patterns of dorsum of *saguinus* group. From P. Hershkovitz (1966). *Folia Primat.* **4**, 381–395.

13

2.2.(1) CALLIMICONIDAE

Only one genus and one species .... 2.2.1.1. *Callimico goeldii*

2.3.(1–11) CEBIDAE

Platyrrhinae larger in size than the Callithricidae with ears more or less naked externally; all digits carrying flattened or curved nails; long, hairy tail (except *Cacajao*)

    (A) Orbits very large; face orthognathous; ears hidden in fur; cheiridia with well-defined, coarsely ridged pads .... 2.3.1.1. *Aotes trivirgatus*

    (B) Orbits normal; their combined width less than cranial length; face more or less prognathous from premaxillary projection; ears visible; cheiridia with normally ridged pattern:

        (a) face short and laterally compressed; canine not projecting .... 2.3.2.(1–3) *Callicebus*

        (b) face never short and compressed; canines long, always projecting beyond incisor level:

            i. lower incisors long, compressed, separated by diastema from canines; canines long, tusk-like, divergent; tail not prehensile .... 2.3.(3–5)

            ii. lower incisors not elongated and not separated from canines by appreciable diastema; tail prehensile:

                (*a*) tail hairy throughout; feet long compared with hands; thumb well developed and not functioning in unison with index..................... 2.3.(7–8)

                (*b*) tail with the distal third naked and ridged beneath; feet short compared with hands; thumb abductor working in unison with index:

                    —hyoid inflated............ 2.3.6.(1–5) *Alouatta*

                    —hyoid not inflated .... 2.3.(9–11)

2.3.2.(1–3) *Callicebus*

   (A) Coat of the back dark brown with long
       buff tips; head with sharply contrasting
       black mask, or similar to back; hands and
       feet black. Tail red–brown .... 2.3.2.1. *C. cupreus*
   (B) Coat of the back reddish–black with
       white, buff or orange throat patch. Feet
       black; hands white. Tail black......... 2.3.2.2. *C. torquatus*
   (C) Coat of the back grey, reddish or brown;
       forehead with grey or black band; hands
       and feet grey, red or dark brown; tail
       dark grey .... 2.3.2.3. *C. gigot*

2.3.(3–5)

   (A) Dolichocephaly pronounced, orbits pro-
       jecting forward separated from the brain
       case in the adult by a pronounced frontal
       depression; nasal swollen above nares;
       absence of long beard beneath chin; tail
       long............................... 2.3.3.(1–2) *Pithecia*
   (B) Dolichocephaly slight, orbits not pro-
       jecting forward; forehead height without
       depression; nasal not swollen above
       nares; incisors and canines abnormal,
       jaws markedly prognathous
       (a) tail long; long beard under chin .... 2.3.4.(1–2) *Chiropotes*
       (b) tail short...................... 2.3.5.(1–3) *Cacajao*

2.3.3.(1–2) *Pithecia*

   (A) Coat thick, coarse and untidy and pre-
       dominantly black; sexual dichromatism
       and dimorphism present. ♂ has black
       muzzle surrounded by creamy–white
       hair; ♀ has lightly-coloured oblique para-
       nasal streaks similar to *P. monachus*...... 2.3.3.1. *P. pithecia*
   (B) Coat colour dark grey and brindled face
       dark with pale-coloured oblique stripes
       on either side of nose .... 2.3.3.2. *P. monachus*

2.3.4.(1–2) *Chiropotes*

   (A) Body with a variable degree of chestnut–
       brown colouration on back, shoulders
       and limbs; face black and naked .... 2.3.4.1. *C. satanas*

(B) Body uniformly black; face with a pink-
ish nose and upper lip .... 2.3.4.2. *C. albinasa*

2.3.5.(1–3) *Cacajao*
    (A) Body fur silvery grey or white; face and
    forehead pink and bare .... 2.3.5.1. *C. calvus*
    (B) Body fur red–brown; face and forehead
    crimson and bare................... 2.3.5.2. *C. rubicundus*
    (C) Body fur chestnut brown with black ex-
    tremities; face and forehead black and
    bare .... 2.3.5.3. *C.
    melanocephalus*

2.3.6.(1–5) *Alouatta*
    (A) Coat black .... 2.3.6.1. *A. villosa*
    (B) Coat brown .... 2.3.6.2. *A. fusca*
    (C) Coat copper–red .... 2.3.6.3. *A. seniculus*
    (D) Coat black but with reddish hands, feet
    and tail tip .... 2.3.6.4. *A. belzebul*
    (E) Coat showing sexual dichromatism: adult
    male black, female and juvenile olive–
    buff............................... 2.3.6.5. *A. caraya*

2.3.(7–8)
    (A) Size small, face covered with short
    whitish hair .... 2.3.7.(1–2) *Saimiri*
    (B) Size medium, face virtually naked except
    for a few bristly hairs .... 2.3.8.(1–4) *Cebus*

2.3.7.(1–2) *Saimiri*
    (A) Crown hair colour from olive green to
    grey–blue .... 2.3.7.1. *S. sciurea*
    (B) Crown hair black................... 2.3.7.2. *S. oerstedii*

2.3.8.(1–4) *Cebus*
    (A) Tufted head composed of long, dark,
    erect hair which may form ridges or
    "horns" with a darkish pre-auricular
    band on the face extending from the cap
    and ending below the chin .... 2.3.8.1. *C. apella*
    (B) Untufted head with a dark, smooth V-
    shaped patch of variable extent; no dark
    pre-auricular band:
        (a) coat black .... 2.3.8.2. *C. capucinus*
        (b) coat light brown or cinnamon ...... 2.3.8.3. *C. albifrons*
        (c) coat light brown or dark brown ..... 2.3.8.4. *C. nigrivittatus*

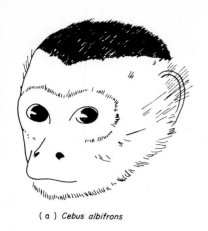

(a) *Cebus albifrons*

(b) *Cebus albifrons* ♀

(c) *Cebus capucinus*

(d) *Cebus nigrivittatus*

(e) *Cebus apella*

(f) *Cebus apella*

Fig. IV. Head patterns of *Cebus*.

Untufted group; (a) *C. albifrons*; (b) *C. albifrons* (♀ with moderately developed superciliary brush); (c) *C. capucinus*; (d) *C. nigrivittatus*.

Tufted group: (e) and (f) *C. apella* with prominent and moderately developed tufts, respectively. From P. Hershkovitz (1949). *Proc. U.S. natn. Mus.* **98**, 323–427.

2.3.(9–11)
- (A) Pollex in the fore-limb absent:
  - (a) limbs extremely lank, hair long and thick . . . . 2.3.9.(1–4) *Ateles*
  - (b) limbs sturdy, hair soft, of woolly texture, pelage dense as in *Lagothrix* . . . . 2.3.10.1. *Brachyteles arachnoides*
- (B) Pollex present in the fore-limb. Coat colour variable from brown or grey to blackish, face black or brownish without buff patch around the nose . . . . 2.3.11.1. *Lagothrix lagotricha*

2.3.9.(1–4) *Ateles*
- (A) Coat black . . . . 2.3.9.1. *A. paniscus*
- (B) Coat black or brownish-black . . . . 2.3.9.2. *A. fusciceps*
- (C) Coat black or brown, generally with paler underpart and a pale triangular patch on the forehead . . . . 2.3.9.3. *A. belzebuth*
- (D) Coat golden, red, buff or dark brown with hands and feet generally black . . . . 2.3.9.4. *A. geoffroyi*

SIMIAE CATARRHINAE: family key

(A) Adult size smaller than man, with or
    without tail .... 3.(1–3)
(B) Adult very large, of size comparable to
    that of man of similar age; tail absent,
    posture only occasionally semi-erect .... 3.4. PONGIDAE
(C) As in (B) but with erect posture .... 3.5. HOMINIDAE

3.(1–3)
(A) Tail short; less than a third of the length
    of the body; in some species apparently
    absent as in *Macaca sylvana* .... 3.1.(1–4)
(B) Tail long: as long as body or longer:
    (a) hand with thumb:
        i. tail without tufted hair; living ex-
            clusively in Africa; male genitalia
            coloured .... 3.1.(5–6)
        ii. tail with or without tufted hair;
            living in East Asia .... 3.2.(1–5)
    (b) hand without thumb; tail with long
        tufted hair; living exclusively in
        Africa .... 3.2.6.(1–5) *Colobus*
(C) Tail absent; very able brachiator....... 3.3. HYLOBATIDAE

3.1.(1–4)
(A) Size moderate to large; body robust; tail
    sometimes long; commonly reduced or
    even absent (as in *M. sylvana*). Living in
    Asia except *M. sylvana*............... 3.1.1.(1–13) *Macaca*
(B) Size very large; body robust; muzzle
    greatly elongated; tail moderate or
    absent:
    (a) muzzle without swellings, merely
        ridges between dorsum and sides; tail
        longer and carried in an arch ...... 3.1.2.(1–5) *Papio*
    (b) muzzle with paired, raised, longitud-
        inal swellings; tail a mere stump .... (6–7) *Papio*
    (c) muzzle shorter, rounded; tail longish,
        tufted at end: naked area on chest .. 3.1.3.1. *Theropithecus
                                              gelada*
(C) Size moderate to large; body slender; tail
    very long. Living in Africa .... 3.1.4.(1–5) *Cercocebus*

19

3.1.1.(1–13) *Macaca*

(A) General fur colour black or very dark brown, sometimes with paler under parts:
    (a) face black and bare
        i. with a large ruff encircling the face, tail long .... 3.1.1.1. *M. silenus*
        ii. with broad flat muzzle owing to longitudinal bony ridges on either side of nose; tail vestigial .... 3.1.1.2. *M. niger*
    (b) face dark flesh-colour, sometimes freckled; forehead and crown covered with erect golden brown hair; tail absent....................... 3.1.1.3. *M. sylvana*
    (c) forehead bare and wrinkled; face pink or red, sometimes with dark freckles .... 3.1.1.4. *M. arctoides*
    (d) face brownish-black and bare; strong brow ridges.................... 3.1.1.5. *M. maura*
(B) General fur colour from dark brown to reddish–brown or gold
    (a) face bare or almost bare
        i. face red, pink or flesh-coloured
            (*a*) general colour gold or reddish–brown; short hair from whorl on cheeks................ 3.1.1.6. *M. sinica*
            (*b*) general colour grey–brown; short hair on forehead growing outwards from centre parting. No whorl on cheeks..... 3.1.1.7. *M. radiata*
            (*c*) general colour brown with darker limbs; dark whiskered cheeks; face relatively short.. 3.1.1.8. *M. cyclopis*
            (*d*) general colour brown; crown hair directed backwards..... 3.1.1.9. *M. mulatta*
            (*e*) general colour yellowish–brown; fur shaggy, tail very short with terminal tuft...... 3.1.1.10. *M. fuscata*
        ii. face light brown, eyelids strongly marked; crown of dark brown erect hair forming a thick cap.... 3.1.1.11. *M. nemestrina*
    (b) face hairy
        i. cheek hair from fringe of whiskers on and around face. Triangle of

20

pale naked skin on inner side of
eyelids...................... 3.1.1.12. *M. fascicularis*
  ii. flesh colour around mouth, darker
yellow eyes. Fringe of darker hair
directed backwards from cheek to
ear......................... 3.1.1.13. *M. assamensis*

3.1.2.(1–5) *Papio*
  (A) Pelage differing in the two sexes, adult♂♂
grey with heavy shoulder mantle, sharply
defined from short hair of croup; adult♀♀
and immatures of both sexes brownish–
olive; face naked, pallid flesh colour.
Tail recurved but not angled. Callosities
large, rosy–red...................... 3.1.2.1. *P. hamadryas*
  (B) Pelage alike in the two sexes; face dusky,
slate grey:
    (a) callosities small, aligned transversely;
body slender; tail sharply angled;
muzzle elongated, its axis at a sharp
angle with that of the brain case. No
cheek tufts...................... 3.1.2.2. *P. ursinus*
    (b) callosities large, occupying the whole
area of the buttocks, rising each side
of base of tail:
      i. size large; back outline more ob-
lique; body thickset; tail markedly
angled. Angle between axis of
muzzle and that of brain case
weak. Apex of muzzle prominent,
even in juveniles, hanging over
base of upper lip. Whiskers dense,
passing gradually into hair of
crown, giving head a spherical as-
pect in face view. Dorsal pelage
passing gradually from crown to
lower back; speckled or even
brindled in some races.......... 3.1.2.3. *P. anubis*
     ii. size large; build slight, slender,
except in old males; limbs elongated. Outline of back only feebly
oblique. Pelage light yellow to
yellowish–brown; whiskers whit-

21

ish; head transversely narrowed. Angle of muzzle and brain case less; tail angled.............. 3.1.2.4. *P. cynocephalus*

iii. size small; silhouette massive; outline of back very oblique. Pelage lengthened over shoulders and sharply defined from short hair of croup. Muzzle–cranial angle well marked, but the former straightening toward its extremity. Whiskers forming dense masses enclosing the face. Tail curved, but not angled..................... 3.1.2.5. *P. papio*

3.1.2.(6–7)

(A) Ears flesh coloured; in ♂ coat is long and thick, dark brown to charcoal grey with fringes of yellow and orange; nose and nostrils lacquer red with longitudinal paranasal swellings of a brilliant electric

P. hamadryas   *P. anubis*   *P. papio*

P. cynocephalus   *P. ursinus*

Fig. V. Diagrammatic representations of the posterior end of the body in species of *Papio*.   *Reproduced by courtesy of W. C. Osman Hill.*

22

blue. The colour combination of the genitalia correspond to the colour combination of the face. The ♀ lacks the bicolour mask of the ♂ . . . . 3.1.2.6. *P. sphinx*

(B) Ears black; coat colour olive green with black mask, beard and cheek tufts white. In the ♂ the perineal area shows a vivid colouration of a somewhat metallic nature. In the ♀ the perineal area is relatively unpigmented . . . . 3.1.2.7. *P. leucophaeus*

3.1.4.(1–5) *Cercocebus*
(A) Coat hair short; mushroom to grey hair with tendency to speckling; no crest on head:
   (a) light mushroom colour. . . . . 3.1.4.1. *C. galeritus*
   (b) dark smoky grey. . . . . 3.1.4.2. *C. atys*
   (c) crown chestnut red . . . . 3.1.4.3. *C. torquatus*
(B) Coat hair long; black hair with no speckling; crest on head:
   (a) crown hair forming a vertical pointed crest like a coconut; eyelids light-coloured. . . . . 3.1.4.4. *C. aterrimus*
   (b) tufts above eyes and an occipital tuft; eyelids dark. . . . . 3.1.4.5. *C. albigena*

3.1.(5–6)
(A) Coat colour predominantly green, yellow, rufous and black, specialized for climbing and jumping on trees able also to run on the ground. Size moderate. . . . 3.1.5.(1–22) *Cercopithecus*

(B) Coat colour red–brown; ground living adapted animals; size large. . . . . 3.1.6.1. *Erythrocebus patas*

3.1.5.(1–22) *Cercopithecus*
(A) Prevailing colour light greenish or greyish . . . . 3.1.5.(1–3)
(B) Prevailing colour green speckled with black on the back, yellowish on arms and legs. Orange rings around eyes. Cheek tufts yellow, radiating in a fan-like

arrangement. Black streak from corner
of eye to ear...................... 3.1.5.22. *C. talapoin*
(C) Prevailing colour not greenish:
   (a) limbs distal to elbows and knees
      wholly black:
      i. Belly blackish; whiskers white... 3.1.5.(6–7)
      ii. Belly whitish; whiskers dark..... 3.1.5.20. *C. mitis*
   (b) Outer part of legs yellow rather than
      black: chin, throat, chest, belly and
      inner part of legs yellowish white.... 3.1.5.21. *C. nigroviridis*
   (c) limbs distal to elbows and knees not
      entirely black:
      i. face black:
         (*a*) with a white brow band and a
            long white beard........... 3.1.5.5. *C. diana*
         (*b*) with chestnut brow band and
            a short white beard......... 3.1.5.15. *C. neglectus*
      ii. face with flesh-coloured muzzle.. 3.1.5.(9–13)
      iii. face with white nose-spot or other
         pattern:
         (*a*) nose-spot cordate, white, yel-
            low or red................ 3.1.5.14. *C. petaurista*
         (*b*) nose-spot heart-shaped rather
            than oval, white........... 3.1.5.(17–19)
         (*c*) nose-spot oval, yellow....... 3.1.5.16. *C. nictitans*
         (*d*) transverse pale blue stripe on
            lip; whiskers yellow......... 3.1.5.4. *C. cephus*
         (*e*) a median vertical white streak
            on nose .... 3.1.5.8. *C. hamlyni*

3.1.5.(1–3) Superspecies *C. aethiops*

(A) Whiskers sharply demarcated in colour
    from the crown; extremities and caudal
    tip light in colour, never black:
   (a) no white frontal band; whiskers
      yellow, ending in a tuft of hairs in
      front of the ear and forming a semi-
      circular crest on temporal region.... 3.1.5.3. *C. sabaeus*
   (b) face sooty black; tuft of white hairs at
      root of tail; scrotum sky blue....... 3.1.5.1. *C. aethiops*
(B) whiskers shorter, not sharply demarcated
    in colour from the crown. Terminal hairs
    of tail black or very dark grey. A patch of
    reddish hairs inferiorly at root of tail and
    around anus...................... 3.1.5.2. *C. cynosuros*

3.1.5.(6–7) Superspecies *C. l'hoesti*
  (A) Cheeks lighter; facial expression benign. 3.1.5.6. *C. l'hoesti*
  (B) Cheeks greyer, strongly speckled with dark grey banding of individual hairs; facial expression fierce................ 3.1.5.7. *C. preussi*

3.1.5.(9–13) Superspecies *C. mona*
  (A) With a narrow line of black hairs on brows (see Fig. VI B)............... 3.1.5.12. *C. pogonias*
  (B) Without sagittal hairy crest on crown;
    (a) distinct dark temporal band extending between eyes and ears (see Fig. VI A) .... 3.1.5.9. *C. mona*
    (b) face bluish, paler around the eyes with lips pink; lower back very dark grey. No white areas on the buttocks 3.1.5.10. *C. campbelli*
    (c) with well developed ear tufts; markedly outstanding white frontal diadem; tail with terminal tuft of elongated hairs .... 3.1.5.11. *C. wolfi*
    (d) similar to *C. mona* and all other species of the group except for the complete absence of darkening on the lower back and hind limbs and for the sharply contrasted white belly. Tail with well defined tuft. .... 3.1.5.13. *C. denti*

3.1.5.(17–19) Superspecies *C. ascanius*
  Nasal spot white and heart-shaped rather than oval.
    (a) face bluish..................... 3.1.5.17. *C. ascanius*
    (b) red hairs on the external ear, tail red and red hairs on the perineal region........................ 3.1.5.18. *C. erythrotis*
    (c) ventral colouration of a faint reddish........................ 3.1.5.19.
                                   C. erythrogaster

3.2.(1–5)
  (A) With the usual monkey nose, scarcely projecting from the rounded muzzle, tail long:
    (a) size moderate to large, tail longer

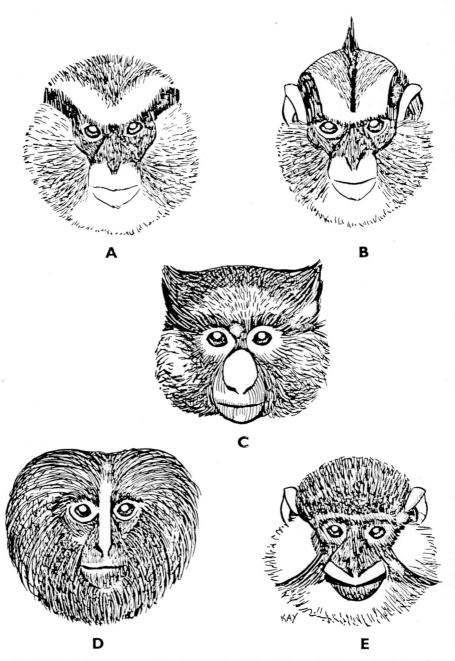

FIG. VI. Facial appearance of various species of *Cercopithecus*. A, *C. mona*; B, *C. pogonias*; C. *C. nictitans*; D. *C. hamlyni*; E, *C. cephus*. From Osman Hill, based on sketches by P. Dandelot.

26

> than body; pelvic limb longer than pectoral . . . . 3.2.1.(1–14) *Presbytis*
>
>   (b) size large; tail shorter than body; pectoral and pelvic limbs subequal . . 3.2.2.1. *Pygathrix nemaeus*

(B) External nose retroussé:

   (a) tail long . . . . 3.2.3.(1–2) *Rhinopithecus*

   (b) tail short . . . . 3.2.4.1. *Simias concolor*

(C) With a large prominent external nose produced into a proboscis . . . . 3 2.5.1. *Nasalis larvatus*

3.2.1.(1–14) *Presbytis*

(A) Natal colour blackish–brown; at 3–5 months changes to light grey; later to brown, slate grey or buff. Brows prominent, brow hair directed forwards and crown hair backwards by a whorl on front part of crown . . . . 3.2.1.1. *P. entellus*

(B) Natal colour grey with white cheeks; adult colour black or grey with a tendency to partial albinism; brown crown hair and whitish whiskers . . . . 3.2.1.2. *P. senex*

   (a) as (B) but black at birth and with long brown crown hair and whiskers in adult . . . . 3.2.1.3. *P. johnii*

(C) Natal colour white with a dark stripe from head to tail tip, extending on to shoulders and forming a cross. With age the dark area gradually extends until only the under parts remain white. Adult colour black, grey or brown . . . . . . 3.2.1.4. *P. aygula* and 3.2.1.5. *P. melalophos*

   (a) entirely white at birth, soon changing to red–brown with bright blue facial skin . . . . 3.2.1.6. *P. rubicundus*

   (b) colour of newborn unknown; a naked white patch low on the forehead . . . . 3.2.1.7. *P. frontatus*

(D) Natal colour bright yellow or orange–
red; coat colour changing to black,
brown or grey by 6 months

    (a) adult black with grey-tipped hair
giving a silver appearance; ♀ with a
patch of paler hair beneath callosi-
ties . . . . 3.2.1.8. *P. cristatus*

    (b) conspicuous white circles round eyes
and white patch on lips. . . . . 3.2.1.9. *P. phayrei*

    (c) conspicuous white circles round eyes
and white patch on lips, but with
hind limbs, tail and crown lighter in
colour than the back. . . . . 3.2.1.10. *P. obscurus*

    (d) adult black with some white on the
head, neck and crest; relatively short
tail . . . . 3.2.1.11. *P. potenziani*

    (e) adult black with variable amount of
white on the head. . . . . 3.2.1.12. *P. francoisi*

    (f) adult grey or blackish-grey with a
thick mat of dark erect hair on the
crown, typically contrasting with the
long whiskers and under parts which
are whitish, buff or rust–red. . . . . 3.2.1.13. *P. pileatus*

    (g) adult cream or gold, the newborn
almost white . . . . 3.2.1.14. *P. geei*

3.2.3.(1–2) *Rhinopithecus*

    (A) Coat colour of back and outer side of
limbs from chocolate to dark grey; forehead, cheeks whiskers and throat variable
from buff to golden–orange to white; tail
from dark yellow–grey to black with tip
variable in colour. . . . . 3.2.3.1. *R. roxellanae*

    (B) Coat colour of back and outer side of
limbs completely black; forehead, cheeks,
whiskers and throat white with orange
throat patch; tail long, dark in colour
with buff tufted tip. . . . . 3.2.3.2. *R. avunculus*

3.2.6.(1–5) *Colobus*

    (A) White mantle and extended fringes; hair
on crown forming an erect cap or
"bonnet" . . . . 3.2.6.2. *C. abyssinicus*

(B) White fur restricted to narrow band, cheeks, neck, shoulders, thighs and tail; "bonnet" absent .... 3.2.6.1. *C. polykomos*

(C) Coat predominantly glossy black with chestnut red forearms, legs and under parts .... 3.2.6.4. *C. badius* and
3.2.6.5. *C. kirkii*

(D) Short haired, olive grey pelage, total lack of adornment. Under parts grey. Tail similar in colour to body.............. 3.2.6.3. *C. verus*

3.3.(1–2) HYLOBATIDAE

(A) Coat colour variable, no web joining the second and third toes; throat covered by hair, no laryngeal sac present......... 3.3.1.(1–6) *Hylobates*

(B) Coat entirely black, with long hair; face and throat naked; second and third toes joined by a web; laryngeal sac present.. 3.3.2.1.
Symphalangus syndactylus

3.3.1.(1–6) *Hylobates*

Coat colour extremely variable as a distinguishing feature:

(A) Crest of erect hair on the crown (elongated at the middle of the crown in ♂, at the side of the crown in ♀); no pale browband present; small throat sac present in ♂; coat colour variable according to age and sex .... 3.3.1.4. *H. concolor*

(B) crown hair directed smoothly backwards:

(a) coat permanently grey–brown with pale brown-band and darker shading on crown and chest.............. 3.3.1.3. *H. moloch*

(b) coat permanently entirely black and coat less dense than the other species. 3.3.1.6. *H. klossii*

(c) coat colour variable according to age and sex; grey at birth, black when adult; with a brown face during puberty in the female; the largest species .... 3.3.1.5. *H. hoolock*

    (d) coat colour variable according to age;
        pale brow-band extended round the
        face to form a complete ring of white
        hair; hands and feet white . . . . 3.3.1.1. *H. lar*

    (e) coat colour variable according to age;
        pale brow-band extended round the
        face to form a complete ring of white
        hair; hands and feet dark (of the same
        colour as the rest of coat) . . . . 3.3.1.2. *H. agilis*

3.4.(1–3) PONGIDAE

(A) Body covered with reddish hair, ears
    small; ♂ with cheek excrescences . . . . 3.4.1.1. *Pongo*
                                       pygmaeus

(B) Body covered by grey–black hair; ears
    small, males with excrescences on vertex;
    hands short and broad; penis short and
    thick . . . . 3.4.2.1. *Gorilla gorilla*

(C) Body covered with glossy black hair;
    ears large, protruding ♂ without facial
    or cranial excrescences; hands long and
    slender; penis long, narrowing to a fine
    apex . . . . 3.4.3.(1–2) *Pan*

3.4.2.(1–2) *Pan*
    (A) Skin colour of face ranging from com-
        pletely white to various degree of black
        pigmentation or of muddy colour; body
        large and robust . . . . 3.4.3.1. *P. troglodytes*

    (B) Skin colour of face completely black;
        stature small (not exceeding 1 m); body
        build paedomorphic, narrow across
        shoulders; body weight less than half that
        of *P. troglodytes*. . . . . 3.4.3.2. *P. paniscus*

3.5.(1) HOMINIDAE

Only one genus and one species . . . . 3.5.1.1. *Homo sapiens*

TABLE OF SUBORDERS, FAMILIES AND GENERA

Suborder	Family	Genus	Number of Described Species
1. PROSIMII	1. TUPAIIDAE	1. *Tupaia*	11
		2. *Dendrogale*	2
		3. *Urogale*	1
		4. *Ptilocercus*	1
	2. LORISIDAE	1. *Loris*	1
		2. *Nycticebus*	2
		3. *Arctocebus*	1
		4. *Perodicticus*	1
	3. GALAGIDAE	1. *Galago*	6
	4. LEMURIDAE	1. *Microcebus*	2
		2. *Cheirogaleus*	2
		3. *Phaner*	1
		4. *Hapalemur*	2
		5. *Lemur*	5
		6. *Lepilemur*	1
	5. INDRIIDAE	1. *Propithecus*	2
		2. *Avahi*	1
		3. *Indri*	1
	6. DAUBENTONIIDAE	1. *Daubentonia*	1
	7. TARSIIDAE	1. *Tarsius*	3
2. SIMIAE PLATYRRHINAE	1. CALLITHRICIDAE	1. *Callithrix*	3
		2. *Cebuella*	1
		3. *Leontideus*	3
		4. *Saguinus*	14
	2. CALLIMICONIDAE	1. *Callimico*	1
	3. CEBIDAE	1. *Aotes*	1
		2. *Callicebus*	3
		3. *Pithecia*	2
		4. *Chiropotes*	2
		5. *Cacajao*	3
		6. *Alouatta*	5
		7. *Saimiri*	2
		8. *Cebus*	4
		9. *Ateles*	4
		10. *Brachyteles*	1
		11. *Lagothrix*	2
3. SIMIAE CATARRHINAE	1. CERCOPITHECIDAE	1. *Macaca*	13
		2. *Papio*	7
		3. *Theropithecus*	1
		4. *Cercocebus*	5
		5. *Cercopithecus*	22
		6. *Erythrocebus*	1
	2. COLOBIDAE	1. *Presbytis*	14
		2. *Pygathrix*	1
		3. *Rhinopithecus*	2
		4. *Simias*	1
		5. *Nasalis*	1
		6. *Colobus*	5

31

Suborder	Family	Genus	Number of Described Species
	3. HYLOBATIDAE	1. *Hylobates*	6
		2. *Symphalangus*	1
	4. PONGIDAE	1. *Pongo*	1
		2. *Pan*	2
		3. *Gorilla*	1
	5. HOMINIDAE	1. *Homo*	1

Tupaia glis, 1.1.1.1.
Tupaia montana, 1.1.1.2.
Tupaia nicobarica, 1.1.1.3.
Tupaia javanica, 1.1.1.4.
Tupaia minor, 1.1.1.5.
Tupaia gracilis, 1.1.1.6.
Tupaia tana, 1.1.1.7.
Tupaia dorsalis, 1.1.1.8.
Tupaia ellioti, 1.1.1.9.
Dendrogale murina, 1.1.2.1.
Dendrogale melanura, 1.1.2.2.
Urogale everetti, 1.1.3.1.
Ptilocercus lowii, 1.1.4.1.
Loris tardigradus, 1.2.1.1.
Nycticebus coucang, 1.2.2.1.
Nycticebus pygmaeus, 1.2.2.2.
Arctocebus calabarensis, 1.2.3.1.
Perodicticus potto, 1.2.4.1.
Galago senegalensis, 1.3.1.1.
Galago crassicaudatus, 1.3.1.2.
Galago alleni, 1.3.1.3.
Galago demidovii, 1.3.1.4.
Galago elegantulus, 1.3.1.5.
Microcebus murinus, 1.4.1.1.
Microcebus coquerelli, 1.4.1.2.
Cheirogaleus major, 1.4.2.1.
Cheirogaleus medius, 1.4.2.2.
Phaner furcifer, 1.4.3.1.
Hapalemur griseus, 1.4.4.1.
Hapalemur simus, 1.4.4.2.
Lemur catta, 1.4.5.1.
Lemur variegatus, 1.4.5.2.
Lemur macaco, 1.4.5.3.
Lemur rubriventer, 1.4.5.4.
Lemur mongoz, 1.4.5.5.
Lepilemur mustelinus, 1.4.6.1.
Propithecus diadema, 1.5.1.1.
Propithecus verreauxi, 1.5.1.2.
Avahi laniger, 1.5.2.1.
Indri indri, 1.5.3.1.
Daubentonia madagascariensis, 1.6.1.1.
Tarsius spectrum, 1.7.1.1.
Tarsius syrichta, 1.7.1.2.
Tarsius bancanus, 1.7.1.3.
Callithrix argentata, 2.1.1.1.

Callithrix humeralifer, 2.1.1.2.
Callithrix jacchus, 2.1.1.3.
Cebuella pygmaea, 2.1.2.1.
Leontideus rosalia, 2.1.3.1.
Leontideus chrysomelas, 2.1.3.2.
Saguinus tamarin, 2.1.4.1.
Saguinus midas, 2.1.4.2.
Saguinus fuscicollis, 2.1.4.3.
Saguinus graellsi, 2.1.4.4.
Saguinus nigricollis, 2.1.4.5.
Saguinus imperator, 2.1.4.6.
Saguinus mystax, 2.1.4.7.
Saguinus labiatus, 2.1.4.8.
Saguinus geoffroyi, 2.1.4.9.
Saguinus oedipus, 2.1.4.10.
Saguinus leucopus, 2.1.4.11.
Saguinus inustus, 2.1.4.12.
Saguinus bicolor, 2.1.4.13.
Saguinus martinsi, 2.1.4.14.
Callimico goeldii, 2.2.1.1.
Aotes trivirgatus, 2.3.1.1.
Callicebus cupreus, 2.3.2.1.
Callicebus torquatus, 2.3.2.2.
Callicebus gigot, 2.3.2.3.
Pithecia pithecia, 2.3.3.1.
Pithecia monachus, 2.3.3.2.
Chiropotes satanas, 2.3.4.1.
Chiropotes albinasa, 2.3.4.2.
Cacajao calvus, 2.3.5.1.
Cacajao rubicundus, 2.3.5.2.
Cacajao melanocephalus, 2.3.5.3.
Alouatta villosa, 2.3.6.1.
Alouatta fusca, 2.3.6.2.
Alouatta seniculus, 2.3.6.3.
Alouatta belzebul, 2.3.6.4.
Alouatta caraja, 2.3.6.5.
Saimiri sciurea, 2.3.7.1.
Saimiri oerstedii, 2.3.7.2.
Cebus apella, 2.3.8.1.
Cebus capucinus, 2.3.8.2.
Cebus albifrons, 2.3.8.3.
Cebus nigrivittatus, 2.3.8.4.
Ateles paniscus, 2.3.9.1.
Ateles fusciceps, 2.3.9.2.
Ateles belzebuth, 2.3.9.3.

Ateles geoffroyi, 2.3.9.4.
Brachyteles arachnoides, 2.3.10.1.
Lagothrix lagotricha, 2.3.11.1.
Macaca silenus, 3.1.1.1.
Macaca nigra, 3.1.1.2.
Macaca sylvana, 3.1.1.3.
Macaca arctoides, 3.1.1.4.
Macaca maura, 3.1.1.5.
Macaca sinica, 3.1.1.6.
Macaca radiata, 3.1.1.7.
Macaca cyclopis, 3.1.1.8.
Macaca mulatta, 3.1.1.9.
Macaca fuscata, 3.1.1.10.
Macaca nemestrina, 3.1.1.11.
Macaca fascicularis, 3.1.1.12.
Macaca assamensis, 3.1.1.13.
Papio hamadryas, 3.1.2.1.
Papio ursinus, 3.1.2.2.
Papio anubis, 3.1.2.3.
Papio papio, 3.1.2.5.
Papio sphinx, 3.1.2.6.
Papio leucophaeus, 3.1.2.7.
Theropithecus gelada, 3.1.3.1.
Cercocebus galeritus, 3.1.4.1.
Cercocebus atys, 3.1.4.2.
Cercocebus torquatus, 3.1.4.3.
Cercocebus aterrimus, 3.1.4.4.
Cercocebus albigena, 3.1.4.5.
Cercopithecus aethiops, 3.1.5.1.
Cercopithecus cynosuros, 3.1.5.2.
Cercopithecus sabaeus, 3.1.5.3.
Cercopithecus cephus, 3.1.5.4.
Cercopithecus diana, 3.1.5.5.
Cercopithecus l'hoesti, 3.1.5.6.
Cercopithecus preussi, 3.1.5.7.
Cercopithecus hamlyni, 3.1.5.8.
Cercopithecus mona, 3.1.5.9.
Cercopithecus campbelli, 3.1.5.10.
Cercopithecus wolfi, 3.1.5.11.
Cercopithecus pogonias, 3.1.5.12.
Cercopithecus denti, 3.1.5.13.
Cercopithecus petaurista, 3.1.5.14.
Cercopithecus neglectus, 3.1.5.15.

Cercopithecus nictitans, 3.1.5.16.
Cercopithecus ascanius, 3.1.5.17.
Cercopithecus erythrotis, 3.1.5.18.
Cercopithecus erythrogaster, 3.1.5.19.
Cercopithecus mitis, 3.1.5.20.
Cercopithecus nigroviridis, 3.1.5.21.
Cercopithecus talapoin, 3.1.5.22.
Erythrocebus patas, 3.1.6.1.
Presbytis entellus, 3.2.1.1.
Presbytis senex, 3.2.1.2.
Presbytis johnii, 3.2.1.3.
Presbytis aygula, 3.2.1.4.
Presbytis melalophos, 3.2.1.5.
Presbytis rubicundus, 3.2.1.6.
Presbytis frontatus, 3.2.1.7.
Presbytis cristatus, 3.2.1.8.
Presbytis phayrei, 3.2.1.9.
Presbytis potenziani, 3.2.1.11.
Presbytis francoisi, 3.2.1.12.
Presbytis pileatus, 3.2.1.13.
Presbytis geei, 3.2.1.14.
Pygathrix nemaeus, 3.2.2.1.
Rhinopithecus roxellanae, 3.2.3.1.
Rhinopithecus avunculus, 3.2.3.2.
Simias concolor, 3.2.4.1.
Nasalis larvatus, 3.2.5.1.
Colobus polykomos, 3.2.6.1.
Colobus abyssinicus, 3.2.6.2.
Colobus verus, 3.2.6.3.
Colobus badius, 3.2.6.4.
Colobus kirkii, 3.2.6.5.
Hylobates lar, 3.3.1.1.
Hylobates agilis, 3.3.1.2.
Hylobates moloch, 3.3.1.3.
Hylobates concolor, 3.3.1.4.
Hylobates hoolock, 3.3.1.5.
Hylobates klossii, 3.3.1.6.
Symphalangus syndactylus, 3.3.2.1.
Pongo pygmaeus, 3.4.1.1.
Pan troglodytes, 3.4.3.1.
Pan paniscus, 3.4.2.2.
Gorilla gorilla, 3.4.3.1.
Homo sapiens, 3.5.1.1.

Reproduced by courtesy of San Diego Zoo

GEOG DIST: South East Asia   BDY WT: 177g♂; 159g♀
HD/BDY LTH: 140–230 mm♂; 143–225 mm♀ TAIL LTH: 129–215 mm;♂130–205 mm♀
COAT COL: brown, speckled blade; shoulder stripe, stomach buff or red-brown

SYNONYMS

Sorex glis Diard 1820
Tupaia ferruginea Raffles 1821
Cladobates belangeri Wagner 1841
Tupaia peguanus Lesson 1842
Sciurus dissimilis Ellis 1860
Tupaia chinensis Anderson 1879
Tupaia picta Thomas 1892
Tupaia phaeura Miller 1902
Tupaia chrysogaster Miller 1903
Tupaia carimatae Miller 1906
Tupaia modesta Allen 1906
Tupaia discolor Lyon 1906
Tupaia concolor Bonhote 1907
Tupaia lacernata Thomas and Wroughton 1909
Tupaia obscura Kloss 1911
Tupaia natunae Lyon 1911
Tupaia belangeri Thomas 1914

35

Tupaia conedor Kloss 1916
Tupaia clarissa Thomas 1917

VERNACULAR NAMES

Spitzhörnchen
Common tree shrew
Tupaïe ferrugineux
Painted tree-shrew
Miller's tree-shrew

TUPAIA MONTANA   Thomas 1892

Reproduced by courtesy of Photo V. Six, Antwerp Zoo

GEOG DIST: South East Asia
HD/BDY LTH: >140 mm
COAT COL: brown, speckled with black

BDY WT: 160g
TAIL LTH: >130 mm

SYNONYMS

Tupaia montana baluensis Lyon 1913

VERNACULAR NAMES

Gebirgsspitzhörnchen von Borneo
Mountain tree-shrew

1.1.1.3. *TUPAIA NICOBARICA* (Zelebor 1869)

GEOG DIST: South East Asia
HD/BDY LTH: <140 mm
COAT COL: brown, speckled with black

BDY WT: 120g
TAIL LTH: >hd/bdy lth

SYNONYMS

Cladobates nicobaricus Zelebor 1869

VERNACULAR NAMES

Nicobar tree-shrew
Toupaïe des Îles Nicobar

1.1.1.4.   *TUPAIA JAVANICA*   Horsfield 1822

GEOG DIST: South East Asia
HD/BDY LTH: <140 mm
COAT COL: brown, speckled with black

BDY WT: 120g
TAIL LTH: >hd/bdy lth

SYNONYMS

VERNACULAR NAMES

Javaspitzhörnchen
Small tree-shrew
Toupaïe de Java

1.1.1.5. *TUPAIA MINOR* Günther 1876

Reproduced by courtesy of M. W. Sorenson (S. Karger AG Basel/New York)

GEOG DIST: South East Asia   BDY WT: 30–60g
HD/BDY LTH: 105–142mm♂;118–170mm♀ TAIL LTH: 142–165 mm♂; 130–165 mm♀
COAT COL: brown, speckled with black

SYNONYMS

Tupaia malaccana Anderson 1879
Tupaia siamensis Gyldenstolpe 1916

VERNACULAR NAMES

Günthers Spitzhörnchen
Günther's tree-shrew
Toupaïe de Müller
Lesser tree-shrew
Midget tree-shrew

1.1.1.6. *TUPAIA GRACILIS* Thomas 1893

GEOG DIST: South East Asia
HD/BDY LTH: <140 mm
COAT COL: brown, speckled with black

BDY WT: 120g
TAIL LTH: >hd/bdy lth

SYNONYMS

VERNACULAR NAMES

Slender tree-shrew

Reproduced by courtesy of F. D'Souza

GEOG DIST: Borneo, Sumatra and　　　　BDY WT: 160–260g
　　smaller islands
HD/BDY LTH: 163–240mm♂; 177–240mm♀　TAIL LTH: 150–196 mm♂; 145–190 mm♀
COAT COL: brown, speckled with black; black stripe on back

SYNONYMS

Cladobates speciosus Wagner 1840
Tupaia sirhassanensis Miller 1901
Tupaia cervicalis Miller 1903
Tana paitana Lyon 1913
Lyonogale tana

VERNACULAR NAMES

Tana
Large tree-shrew
Terrestrial tree-shrew

1.1.1.8. *TUPAIA DORSALIS* Schlegel 1875

GEOG DIST: Borneo BDY WT: 105g
HD/BDY LTH: 161–220 mm♂; 175–210 mm♀ TAIL LTH: 145–150 mm♂; 140–150 mm♀
COAT COL: brown, speckled with black; black stripe to root of tail

SYNONYMS

VERNACULAR NAMES

Striped tree-shrew

1.1.1.9. *TUPAIA ELLIOTI* Waterhouse 1850

GEOG DIST: South East Asia
HD/BDY LTH: *ca* 150–200mm
COAT COL: red-brown or grey-brown, speckled with black; large hairy ears

BDY WT: *ca* 150g
TAIL LTH: < hd/bdy lth

SYNONYMS

Anathana wroughtoni Lyon 1913
Anathana pallida Lyon 1913

VERNACULAR NAMES

Indisches Spitzhörnchen
Madras tree-shrew
Toupaïe d'Elliot

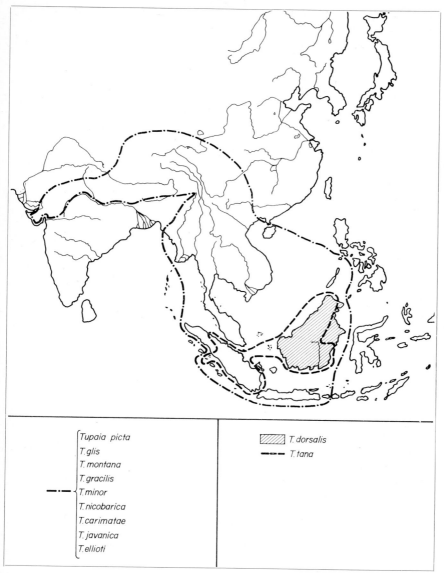

Tupaia picta	
T. glis	
T. montana	
T. gracilis	
T. minor	▨ T. dorsalis
T. nicobarica	— ■ ■ — T. tana
T. carimatae	
T. javanica	
T. ellioti	

MAP 1. Geographic distribution of the different species of *Tupaia* (redrawn and adapted from Fiedler, 1956 and Napier, 1967)

45

1.1.2.1. *DENDROGALE MURINA* (Schlegel and Müller 1845)

GEOG DIST: South Vietnam, South East BDY WT: *ca* 100g
  Thailand, Cambodia
HD/BDY LTH: 103–130 mm♂; 107–150 mm♀ TAIL LTH: 110–145 mm♂; 105–145 mm♀
COAT COL: grey-brown, lighter under parts; face with dark streak

SYNONYMS

Hylegalea murina Schlegel and Müller 1845
Tupaia frenata Gray 1860

VERNACULAR NAMES

Northern smooth-tailed tree-shrew
Toupaïe murin

1.1.2.2. *DENDROGALE MELANURA* (Thomas 1892)

GEOG DIST: North East Borneo   BDY WT: *ca* 100g

HD/BDY LTH: 103–130mm♂; 107–150mm♀ TAIL LTH: 110–145 mm♂; 105–145 mm♀

COAT COL: dark grey, orange on cheeks and around eyes

SYNONYMS

VERNACULAR NAMES

Southern smooth-tailed tree-shrew

1.1.3.1.   *UROGALE EVERETTI*   (Thomas 1892)

Reproduced by courtesy of R. Buchsbaum

GEOG DIST: Philippines   BDY WT: 355g♂
HD/BDY LTH: 182–235 mm♂; 200–202 mm♀ TAIL LTH: 148–170 mm♂; 147–150 mm♀
COAT COL: dark brown, yellow or rufous under parts

SYNONYMS

Urogale cylindrura Mearns 1905 von Mindanao

VERNACULAR NAMES

Everett's tupaia
Philippinenspitzhörnchen
Philippine tree-shrew

48

1.1.4.1. *PTILOCERCUS LOWII* (Gray 1848)

Drawing by Kostas, Turin

GEOG DIST: South East Asia   BDY WT: 469g♂; 304g♀
HD/BDY LTH: 120–143 mm♂; 120–140 mm♀ TAIL LTH: 165–180 mm♂; 170–180 mm♀
COAT COL: grey; tail scaly, dark, with terminal portion tufted like a feather

SYNONYMS

VERNACULAR NAMES

Fahnenschwanzspitzhörnchen
Feather-tailed tree-shrew
Pen-tailed tree-shrew
Treemouse
Ptilocerque de Low

Reproduced by courtesy of Oregon Regional Primate Research Center

GEOG DIST: Ceylon, South India   BDY WT: 85–348g♂; 85–270g♀
HD/BDY LTH: 186–264g♂; 198–243g♀   TAIL LTH: no tail
COAT COL: pale grey, brown or red-brown; white or buff under parts

1.2.1.1.   *LORIS TARDIGRADUS*   (Linnaeus 1758)

SYNONYMS

Lemur tardigradus L. 1758
Stenops gracilis Geoffroy 1796
Loris gracilis Geoffroy 1796
Nycticebus gracilis Blainville 1841
Stenops tardigradus V.D. Hoeren 1844

VERNACULAR NAMES

Schlankloris
Slender loris
Loris grêle
Sherminda
Thevangu

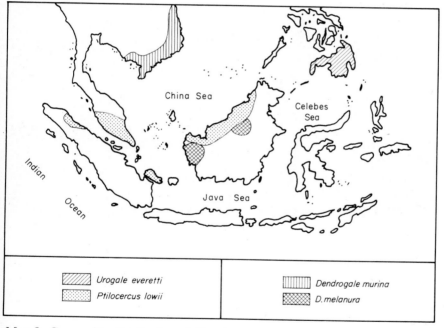

MAP 2. Geographic distribution of *Urogale*, *Ptilocercus* and the different species of *Dendrogale* (redrawn and adapted from Fiedler, 1956)

1.2.2.1. *NYCTICEBUS COUCANG* (Boddaert 1785)

GEOG DIST: South East Asia BDY WT: 1012–1675g♂; 1105–1370g♀
HD/BDY LTH: 265–280mm♂; 268–335mm♀ TAIL LTH: very short
COAT COL: brown, red-brown or grey; white between eyes, dark circles round eyes, stripe (medial) on crown

SYNONYMS

Tardigradus coucang Boddaert 1783
Lemur tardigradus Link 1795

VERNACULAR NAMES

Plumplori Kukang
Slow loris Oukang
Loris paresseux

1.2.2.2.   *NYCTICEBUS PYGMAEUS*   Bonhote 1907

GEOG DIST: South East Asia
HD/BDY LTH: *ca* 190mm
COAT COL: brown, red-brown or grey

BDY WT: *ca* 500g
TAIL LTH: very short

SYNONYMS

VERNACULAR NAMES

Kleiner Plumplori
Lesser slow loris
Pygmy slow loris

1.2.3.1. *ARCTOCEBUS CALABARENSIS* (Smith 1860)

Reproduced by courtesy of Oregon Regional Primate Research Center

GEOG DIST: Tropical forests of West   BDY WT: 266–465g
   Africa
HD/BDY LTH: 220–251 mm♂; 231–263 mm♀ TAIL LTH: *ca* 8 mm
COAT COL: rust or yellow-brown above; fawn or pale grey on underparts

1.2.3.1. *ARCTOCEBUS CALABARENSIS* (Smith 1860)

SYNONYMS

Stenops calabarensis Smith 1860
Perodicticus calabarensis Smith 1860

VERNACULAR NAMES

Bärenmaki
Angwántibo
Golden potto
Pérodictique angwantibo
Potto de Calabar
Potto doré
Dwan

1.2.4.1.   *PERODICTICUS POTTO*   (Müller 1766)

GEOG DIST: Central, West Africa   BDY WT: 1025–1400g♂; 1000–1200g♀
HD/BDY LTH: 337–406mm♂; 355–417mm♀   TAIL LTH: 50–81 mm♂; 56–72 mm♀
COAT COL: variable; grey to brown, lighter underparts

SYNONYMS

Lemur potto P. S. Müller 1766
Nycticebus potto Geoffroy 1812
Galago guineensis Desmarest 1820
Perodycticus geoffroyi Bennet 1839
Stenops potto van der Hoeven 1847–49 and Pel 1852
Perodicticus edwardsi Bouvier 1879
Perodicticus faustus Thomas 1910

1.2.4.1. *PERODICTICUS POTTO* (Müller 1766)

Potto
Pérodictique potto
Half-a-tail
Aposou
Aposo
Awun
Iki

Orunoleu
Kikami
Shakami
Kabende

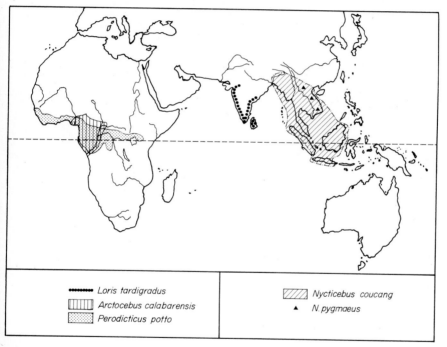

MAP 3. Geographic distribution of *Loris*, *Arctocebus*, *Perodicticus* and the different species of *Nycticebus* (redrawn and adapted from Fiedler, 1956)

57

1.3.1.1. *GALAGO SENEGALENSIS* E. Geoffroy 1796

Reproduced by courtesy of Tierbilder Okapia

GEOG DIST: Uganda, Kenya, Tanzania,
 Zambia, Senegal, Guinea, Ethiopia   BDY WT: 300g♂; 229g♀
HD/BDY LTH: 151–173 mm♂; 140–163 mm♀   TAIL LTH: 217–250 mm♂; 205–248 mm♀
COAT COL: grey to red-brown

SYNONYMS

Otolicnus teng Sundeval 1842
Galago murinus Murrey 1859
Galago sennariensis Gray 1863
Galago zansibaricus Matschie 1893
Galago gallarum Thomas 1901
Euoticus inustus Schwarz 1930

1.3.1.1. *GALAGO SENEGALENSIS* E. Geoffroy 1796

Senegal galago
Zwerggalago
Lesser galago
Bush-baby
Lesser bush-baby
Galago commun
Lende
Ngei
Katinda mugogo

Zanzibar galago
Komba kidogo
Kideri
Komba
Idekelwa
Adokole
Okodu
Kamenamena

1.3.1.2. *GALAGO CRASSICAUDATUS* E. Geoffroy 1812

Reproduced by courtesy of Tierbilder Okapia

GEOG DIST: South Central Africa, Angola   BDY WT: 1240g♂; 1035g♀
HD/BDY LTH: 320–375mm♂; 295–335mm♀ TAIL LTH: 440–475 mm♂;415–425 mm♀
COAT COL: grey to red-brown; outside of limbs rust

1.3.1.2. *GALAGO CRASSICAUDATUS* E. Geoffroy 1812

SYNONYMS

Otolicnus garnettii Ogilby 1838
Otolemur agisymbanus Coquerell 1859
Callotus monteiri Gray 1863
Otolemur badius Matschie 1905

VERNACULAR NAMES

Riesengalago
Grand galago
Thick-tailed galago
Galage à queue touffue
Large grey night ape
Great bush-baby
Le grand galago
Thicktailed bush-baby
Grey lemur
Bosaap
Rat of the coconut palm
Suikiui
Garila
Greater galago
Komba
Kulo adadi

Reproduced by courtesy of Zoological Society of London

GEOG DIST: Tropical forest of Cameroun,   BDY WT: *ca* 250g
    Gabon
HD/BDY LTH: *ca* 150 mm   TAIL LTH: *ca* 230 mm
COAT COL: grey to red-brown, yellowish underparts; outside of limbs rust tinged

SYNONYMS

Galago cameronensis Peters 1876
Galago gabonensis Elliot 1907
Galago batesi Elliot 1907

VERNACULAR NAMES

Galago d'Allen
Black-tailed bush-baby
Allen's galago

GALAGO DEMIDOVII (Fischer 1808)

Reproduced by courtesy of Zoological Society of London

GEOG DIST: Central Africa
HD/BDY LTH: 125–160 mm
BDY WT: 50–80g
TAIL LTH: 150–215 mm
COAT COL: variable; brown to yellow-brown; yellowish underparts, pale streak down nose

1.3.1.4.   *GALAGO DEMIDOVII*   (Fischer 1808)

SYNONYMS

Otolicnus peli Temminck 1853
Hemigalago demidoffi Dahlhon 1857
Galago murinus Murray 1859
Otolicnus pusillus Peters 1876

VERNACULAR NAMES

Demidoff-galago
Demidoff's bush-baby
Dwarf galago
Galago de Demidoff
Dwarf bush-baby
Pigmy galago
Endinda
Kanduku

1.3.1.5.   *GALAGO ELEGANTULUS*   (Leconte 1857)

Reproduced by courtesy of Zoologischer Garten, Leipzig

GEOG DIST: West forests between Congo and Cross Rivers

BDY WT: *ca* 250–330g

HD/BDY LTH: 187–235 mm

TAIL LTH: 280–332 mm

COAT COL: cinnamon, dark dorsal median stripe; pale grey beneath

SYNONYMS

Microcebus elegantulus Leconte 1857
Otolicnus apicalis Du Chaillu 1860
Euoticus elegantulus Gray 1863
Otolicnus tonsor Dollman 1910

VERNACULAR NAMES

Galago mignon
Needle-nailed galago

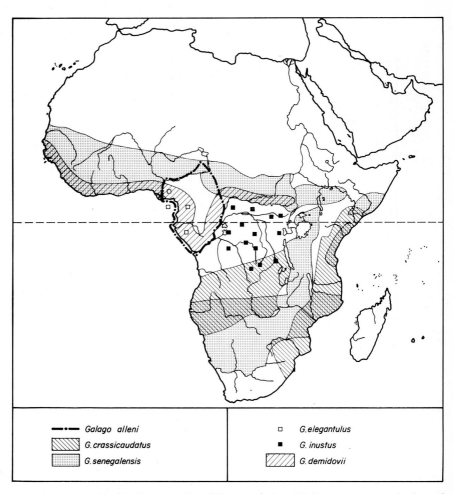

MAP 4. Geographic distribution of the different species of *Galago* (redrawn and adapted from Hill, 1953)

1.4.1.1. *MICROCEBUS MURINUS* (Miller 1777)

Reproduced by courtesy of Zoological Society of London

GEOG DIST: Madagascar
HD/BDY LTH: *ca* 130 mm
COAT COL: brown or grey; white median nasal stripe

BDY WT: *ca* 60g
TAIL LTH: *ca* 170 mm

SYNONYMS

Prosimia minima (Boddaert 1785)
Lemur prehensilis Kerr 1792
Lemur pusillus Geoffroy 1795
Galago madagascariensis Geoffroy 1812
Cheirogaleus minor Geoffroy 1812
Microcebus rufus Wagner 1840
Microcebus palmarum Lesson 1840
Microcebus myoxinus Peters 1852
Microcebus gliroides Grandidier 1868
Microcebus nain Duncan 1883
Microcebus griseorufus Kollmann 1910

1.4.1.1.   *MICROCEBUS MURINUS*   (Miller 1777)

Mausmaki
Microcèbe
Miller's mouse-lemur
Chirogale mignon
Lesser mouse lemur
Tily
Taidy
Kely-be-ohy

1.4.1.2. *MICROCEBUS COQUERELLI* (Grandidier 1867)

Reproduced by courtesy of Zoological Society of London

Geog Dist: West Madagascar
Hd/Bdy Lth: *ca* 250 mm
Coat Col: brown or grey; no nasal stripe

Bdy Wt: *ca* 350g
Tail Lth: *ca* 280 mm

SYNONYMS

Cheirogaleus coquerelli Grandidier 1867
Mirza coquerelli Gray 1870

VERNACULAR NAMES

Coquerell's Katzenmaki
Coquerell's mouse-lemur
Microcèbe de Coquerell
Coquerel's dwarf lemur
Sisiba
Sietui
Setohy
Tsitsihy

1.4.2.1.   *CHEIROGALEUS MAJOR*   E. Geoffroy 1812

GEOG DIST: East Madagascar
HD/BDY LTH: 190–270 mm
COAT COL: brown-red or grey; lighter under parts

BDY WT: 350g
TAIL LTH: 165–250 mm

SYNONYMS

Lemur commersonii Wolf 1822
Cheirogaleus milii Geoffroy 1828
Myspithecus typus Cuvier 1833
Mioxicebus griseus Lesson 1840
Chirogalus crossleyi Grandidier 1870
Chirogale melanotis Forsyth-Major 1894
Chirogale sibreei Forsyth-Major 1896

VERNACULAR NAMES

Milius's Katzenmaki
Greater dwarf-lemur
Chirogale de Milius

1.4.2.2. *CHEIROGALEUS MEDIUS* E. Geoffroy 1812

Reproduced by courtesy of J. J. Petter

GEOG DIST: West and South Madagascar   BDY WT: 250g
HD/BDY LTH: 190–265 mm   TAIL LTH: 165–250 mm
COAT COL: brown-red or grey; white underparts

SYNONYMS

Chirogalus samati Grandidier 1868
Opolemur thomasi Forsyth-Major 1894
Opolemur samati Forsyth-Major 1894
Microcebus samati Kollmann 1910
Altililemur medius Elliot 1913

VERNACULAR NAMES

Fettschwanzmaki
Fat-tailed lemur
Dwarf-lemur
Fat-tailed dwarf lemur
Kely behoy

71

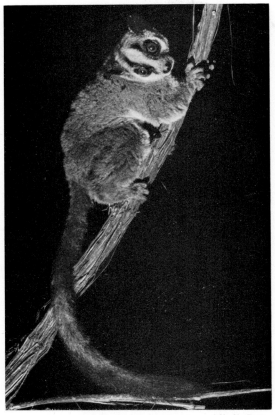

Reproduced by courtesy of J. J. Petter

GEOG DIST: N and W Madagascar   BDY WT: 350g
HD/BDY LTH: 250–275 mm   TAIL LTH: 325–350 mm
COAT COL: light red under parts; dark spinal stripe, bifurcating on crown

SYNONYMS

Microcebus furcifer
Cheirogaleus furcifer

VERNACULAR NAMES

Gabelstreifiger Zwergmaki
Fork-tailed mouse lemur
Phaner à fourche
Fork-crowned dwarf-lemur

72

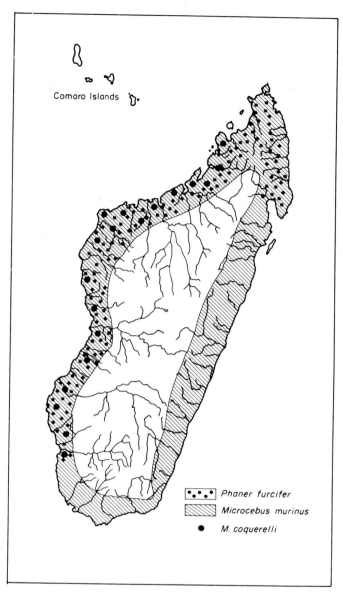

Comoro Islands

.•.•. Phaner furcifer
Microcebus murinus
● M. coquerelli

Map 5. Geographic distribution of *Phaner* and the different species of *Microcebus* (re-drawn and adapted from Hill, 1953)

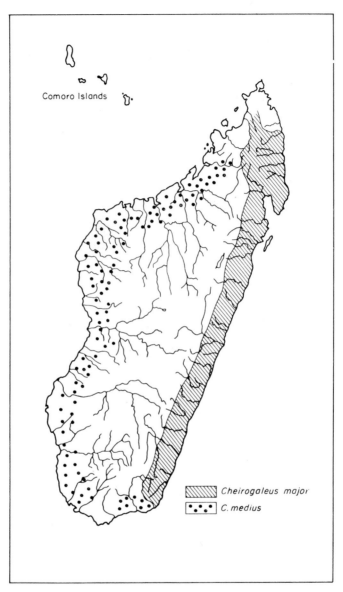

MAP 6. Geographic distribution of the different species of *Cheirogaleus* (redrawn and adapted from Hill, 1953)

1.4.4.1.   *HAPALEMUR GRISEUS*   (Link 1795)

Reproduced by courtesy of W. C. Osman Hill

GEOG DIST: East Madagascar
HD/BDY LTH: 365–370g♂; 260–330g♀
BDY WT: *ca* 1000g
TAIL LTH: 365 mm♂; 240–350 mm♀
COAT COL: green-grey, washed with red, speckled black; limbs and tail soot grey

75

1.4.4.1. *HAPALEMUR GRISEUS* Link 1795

SYNONYMS

Lemur griseus Link 1795, E. Geoffroy 1796
Lemur cinereus Desmarest 1820
Hapalemur olivaceus Geoffroy 1851
Cheirogaleus griseus Giebel 1856
Myoxicebus griseus Elliot 1913

VERNACULAR NAMES

Grauer Halbmaki
Hapalémur gris
Bokombouli
Grey lemur

HAPALEMUR SIMUS Gray 1870

Reproduced by courtesy of Paul Popper

GEOG DIST: East Madagascar
HD/BDY LTH: 300–450 mm
COAT COL: Grey-brown

BDY WT: *ca* 2500g
TAIL LTH: 350–560 mm

SYNONYMS

Prolemur simus Gray 1870 and Pocock 1917

VERNACULAR NAMES

Breitschnauziger Halbmaki
Hapalémur à nez large
Red lemur
Bandro
Broad-nosed lemur
Snub-nosed lemur

Reproduced by courtesy of Zoologischer Garten, Zürich

GEOG DIST: South Madagascar   BDT WT: 2100g
HD/BDY LTH: 303–456 mm   TAIL LTH: 370–560 mm
COAT COL: grey on back; lighter grey limbs, extremities white; tail ringed black/white

1.4.5.1.   *LEMUR CATTA*   Linnaeus 1758

SYNONYMS

Prosimia catta Boddaert 1784
Maki macoco Muirhead 1819
Lemur macoco F. Cuvier 1824

VERNACULAR NAMES

Katzenmaki
Ring-tailed lemur
Lémur catta
Cat lemur
Mac
Varikia
Gidro
Hira

Reproduced by courtesy of W. C. Osman Hill

GEOG DIST: East Madagascar   BDY WT: *ca* 3000 g
HD/BDY LTH: *ca* 600 mm   TAIL LTH: *ca* 600 mm
COAT COL: variegated black/white, or black/red/white

SYNONYMS

Lemur ruber E. Geoffroy 1812
Lemur vari Muirhead 1819, Coquerel 1859
Lemur erythromela Lesson 1840
Lemur varius I. Geoffroy 1851
Varecia variegata Petter 1965

VERNACULAR NAMES

Lémur vari
Ruffed lemur
Varecia
Varimena
Varikandra

80

LEMUR MACACO   Linnaeus 1766

Reproduced by courtesy of Tierbilder Okapia

GEOG DIST: N.W. Madagascar
HD/BDY LTH: 300–450 mm
COAT COL:♂black, ♀brown or reddish

BDY WT: *ca* 2000g
TAIL LTH: 350–560 mm

LEMUR MACACO

SYNONYMS

Lemur niger Schreber 1775
Lemur albifrons E. Geoffroy 1796
Prosimia xanthomystax Gray 1863
Lemur rufifrons Bennet 1833
Lemur fulvus Geoffroy 1812
Lemur bruneus V. D. Hoeven 1844
Lemur leucomystax Bartlett 1862

VERNACULAR NAMES

Mohrenmaki
Black lemur
Lémur macoco
Schwaezhopfmaki
Brown lemur
Lémur brun
Fulvous lemur
Acoumba
Akomba

LEMUR RUBRIVENTER   I. Geoffroy 1850

Reproduced by courtesy of Tierbilder Okapia

GEOG DIST: Madagascar
HD/BDY LTH: 300–450 mm
COAT COL: red-brown, rufous limbs

BDY WT: *ca* 2000g
TAIL LTH: 350–560 mm

SYNONYMS

Lemur flaviventer I. Geoffroy 1850
Lemur rufiventer Gray 1870
Prosimia rufipes Gray 1871

VERNACULAR NAMES

Rotbauchmaki
Red-bellied lemur
Lémur à ventre rouge
Soaemeria

LEMUR MONGOZ Linnaeus 1766

Reproduced by courtesy of Zoologischer Garten, Basel

GEOG DIST: N. and N.W. Madagascar   BDY WT: *ca* 2000g
   Comoro Is.
HD/BDY LTH: 300–450 mm   TAIL LTH: 350–560 mm
COAT COL: dark grey-brown; reddish tinged cheeks and under parts

SYNONYMS

Lemur nigrifrons E. Geoffroy 1812
Lemur albimanus E. Geoffroy 1812
Lemur dubius F. Cuvier 1834
Prosimia micromongoz Lesson 1840
Prosimia bugi Lesson 1840
Prosimia ocularis Lesson 1840
Lemur coronatus Gray 1842
Lemur cuvieri Fitzinger 1870
Prosimia brissonianus Gray 1870
Lemur anjuanensis Peters 1879
Lemur johannae Trouessart 1904
Lemur mongoz Elliot 1913

VERNACULAR NAMES

Mongozmaki   Akomba
Mongoose lemur   Ankomba
Lémur mongos   Gidro

1.4.6.1. *LEPILEMUR MUSTELINUS* I. Geoffroy 1851

Reproduced by courtesy of R. D. Martin

GEOG DIST: Madagascar
HD/BDY LTH: 280–356 mm
BDY WT: *ca* 300g
TAIL LTH: 254–280 mm
COAT COL: red-brown or grey; underparts and/or hind limbs pale grey or yellow-white

SYNONYMS

Lepilemur ruficaudatus Grandidier 1867
Lepilemur dorsalis Gray 1870
Lepilemur pallidacauda Gray 1872
Mixocebus caniceps Peters 1875
Lepilemur microdon Forsyth-Major 1894
Lepilemur edwardsi Forsyth-Major 1894
Lepilemur grandidieri Forsyth-Major 1894
Lepilemur leucopus Forsyth-Major 1894
Lepidolemur globiceps Forsyth-Major 1894
Lepilemur rufescens Lorenz-Liburnau 1898

VERNACULAR NAMES

Wieselmaki
Weasel-lemur
Lépilémur mustélin
Hattock
Sportive lemur
Rotschwanziger Wieselmaki

Red-tailed sportive lemur
Lépilémur à queue rouge
Fitili-ki
Repahaka
Bohenghé

85

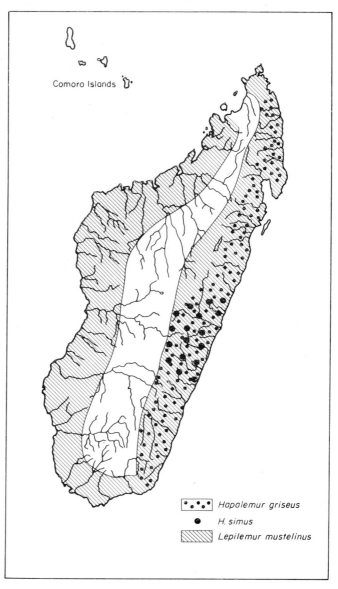

Map 7. Geographic distribution of *Lepilemur* and the different species of *Hapalemur* (redrawn and adapted from Hill, 1953)

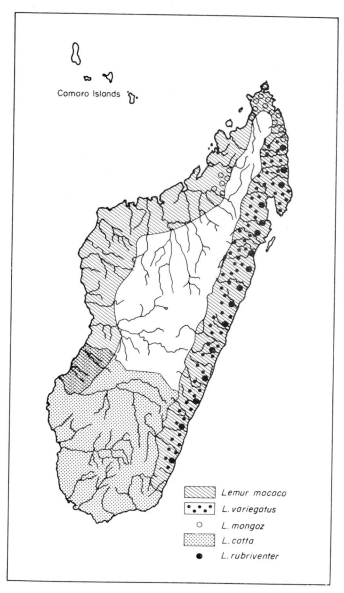

Comoro Islands

Lemur macaco	
L. variegatus	
L. mongoz	
L. catta	
L. rubriventer	

MAP 8. Geographic distribution of the different species of *Lemur* (redrawn and adapted from Hill, 1953)

1.5.1.1. *PROPITHECUS DIADEMA* Bennet 1832

Reproduced by courtesy of J. J. Petter

GEOG DIST: East Madagascar   BDY WT: 6000g
HD/BDY LTH: 458–534 mm   TAIL LTH: 483–560 mm
COAT COL: variable, but predominantly white or pale grey

SYNONYMS

Macromerus typicus A. Smith 1833
Lemur diadema Blainville 1839
Habrocebus diadema Wagner 1840
Indris albus Vinson 1862
Propithecus candidus Grandidier 1871
Propithecus edwardsi Grandidier 1871

VERNACULAR NAMES

Diadem-sifaka   Uliessmaki
Diademed sifaka   Indri
Propithèque diadème   Simpoune
Sifaka   Chimpo
Simpona

1.5.1.2. *PROPITHECUS VERREAUXI* Grandidier 1867

GEOG DIST: West Madagascar

HD/BDY LTH: 460–535 mm

COAT COL: variable, but predominantly pale grey or white

BDY WT: 4000–5000g

TAIL LTH: 485–560 mm

1.5.1.2. *PROPITHECUS VERREAUXI* Grandidier 1867

Propithecus coronatus Milne-Edwards 1871
Propithecus majori W. Rothschild 1894

VERNACULAR NAMES

Verreaux's Sifaka
Propithèque de Verreaux
Sifaka
Simpona
Chimpo
Vliessmaki
Simpoune

AVAHI LANIGER (Gmelin 1788)

Reproduced by courtesy of R. D. Martin

GEOG DIST: North West and East
   Madagascar
HD/BDY LTH: *ca* 320 mm
COAT COL: uniformly brown

BDY WT: 1000g
TAIL LTH: *ca* 390 mm

SYNONYMS

Indri longicaudatus E. Geoffroy 1756
Lemur laniger Gmelin 1788
Lemur brunneus Link 1795
Lichanotus laniger Illiger 1811
Indris longicaudatus E. Geoffroy 1812
Avahis laniger I. Geoffroy 1851; Lesson 1838
Lichanotus awahi van der Hoeven 1844

1.5.2.1.   *AVAHI LANIGER*   (Gmelin 1788)

Wollmaki
Woolly lemur
Avahi lanigère
Woolly avahi
Flocky lemur
Fotsiefaka
Fotsi-fé
Amponghi
Avahy

INDRI INDRI Gmelin 1788

Reproduced by courtesy of J. J. Petter

GEOG DIST: East Madagascar   BDY WT: 7000g
HD/BDY LTH: *ca* 700 mm   TAIL LTH: *ca* 30 mm
COAT COL: variable, predominantly black and white

SYNONYMS

Indris brevicaudatus E. Geoffroy and Couvier 1796
Indris niger Lacépède 1800
Indris ater I. Geoffroy 1825
Indris mitratus Peters 1871
Indris variegatus Gray 1872

VERNACULAR NAMES

Short-tailed indris   Indris
Babakoto   Amboanala
Dog-Faced lemur   Indri
Endrina

93

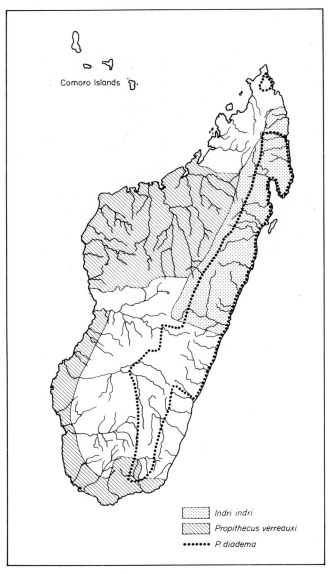

MAP 9. Geographic distribution of *Indri* and the different species of *Propithecus* (redrawn and adapted from Hill, 1953)

1.6.1.1.
DAUBENTONIA MADAGASCARIENSIS (Gmelin 1788)

Reproduced by courtesy of J. J. Petter

GEOG DIST: N.E. Madagascar
HD/BDY LTH: *ca* 400 mm
BDY WT: 2000g
TAIL LTH: 560–600 mm
COAT COL: dark brown to black; face and underparts lighter

SYNONYMS

Sciurus madagascariensis Gmelin 1788
Lemur psilodactylus Shaw 1800
Tarsius daubentonii Shaw 1800
Chiromys madagascariensis E. Geoffroy 1803
Otolicnus madagascariensis van der Hoeven 1803
Daubentonia robusta Lamberton 1934
Chiromys robustus Lamberton 1934

VERNACULAR NAMES

Aye-aye
Hai-hai
Ahay
Aiay
Fingertier

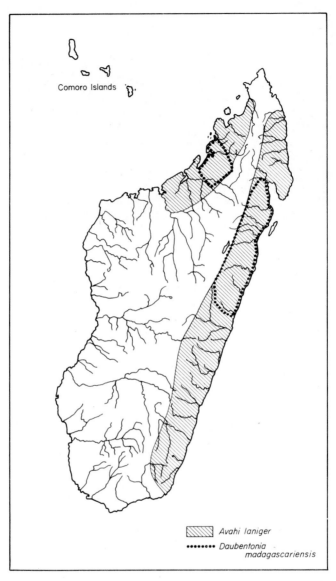

MAP 10. Geographic distribution of *Avahi* and *Daubentonia* (redrawn and adapted from Hill, 1953)

1.7.1.1.   *TARSIUS SPECTRUM*   (Pallas 1778)

Reproduced by courtesy of Paul Popper

Geog Dist: Celebes Is.

Hd/Bdy Lth: 84–160 mm♂; 95–160 mm♀

Coat Col: dark grey-brown; white postauricular spot

Bdy Wt: 95–165g♂; 85–155g♀

Tail Lth: 135–275 mm♂; 190–240 mm♀

SYNONYMS

Lemur podje Kerr 1792
Tarsius pallassii Fischer 1804
Tarsius fuscus Fischer 1804
Tarsius fuscomanus Fischer 1804
Tarsius fischeri Bormeister 1846
Tarsius sangirensis Meyer 1897

97

1.7.1.1. *TARSIUS SPECTRUM* (Pallas 1778)

Östlicher Koboldmaki
Palla's tarsier
Spectral tarsier
Tarsier spectre
Celebesian tarsier
Dusky-handed tarsier
Yellow-bearded tarsier
Celebes tarsier
Eastern tarsier
Podje
Pluimstaarte-Spookdiertje

1.7.1.2.   *TARSIUS SYRICHTA*   (Linnaeus 1758)

Reproduced by courtesy of Paul Popper

Geog Dist: Philippines
Hd/Bdy Lth: 85–160 mm♂; 95–160m m♀
Coat Col: grey tinged with red-brown

Bdy Wt: 95–165g♂; 85–155g♀
Tail Lth: 135–275 mm♂; 190–240 mm♀

SYNONYMS

Yerboa tarsier Zimmermann 1777
Tarsius tarsier Erxleben 1777
Didelphys macrotarsos Schreber 1778
Lemur podje Kerr 1792
Tarsier macauco Pennent 1793
Macrotarsus buffonii Link 1795
Tarsius daubentonii Audebert 1797, 1800
Lemur tarsius Cuvier 1798
Tarsius spectrum E. Geoffroy 1812
Tarsius tarsius Forbes 1894

99

1.7.1.2.   *TARSIUS SYRICHTA*   (Linnaeus 1758)

Tarsius philippensis Meyer 1894–95
Tarsius philippinensis F. Major 1901; Meyer 1896
Tarsius fraterculus Miller 1910
Tarsius carbonarius Heude 1898
Cercopithecus luzonis minimus Camel 1705

VERNACULAR NAMES

Philippinen-Koboldmaki
Buffon's tarsier
Tarsier des Philippines
Philippine tarsier
Amus
Umus
Malmag
Mago
Magau
Magatilakok
Maomag
Mal

1.7.1.3.   *TARSIUS BANCANUS*   Horsfield 1821

Reproduced by courtesy of Paul Popper

GEOG DIST: South East Sumatra, Borneo
HD/BDY LTH: 85–160 mm♂; 95–160 mm♀
COAT COL: grey tinged with golden brown

BDY WT: 95–165g♂; 90–155g♀
TAIL LTH: 135–275 mm♂; 190–240 mm♀

SYNONYMS

Lemur tarsier Raffles 1822
Mypsicebus bancanus Lesson 1840
Tarsius spectrum Weber 1893
Tarsius tarsier Lyon 1906

Tarsius borneanus Elliot 1910
Tarsius saltator Elliot 1910
Tarsius natunensis Chasen 1940

VERNACULAR NAMES

Westlicher Koboldmaki
Horsfield's tarsier
Malaysian tarsier
Raffles tarsier
Western tarsier
Singapooa
Singapuar
Krabuku
Kabuku

Imbing
Tempiling
Perock-poear
Palele
Tindok rokok
Linpseng
Sempalili
Lakud

101

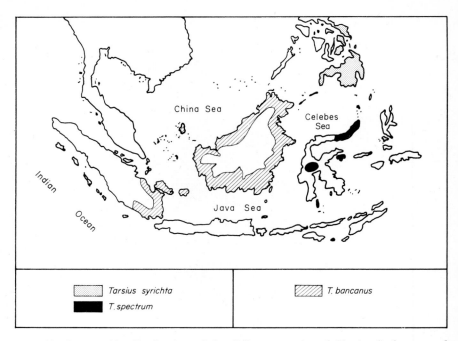

Map 11. Geographic distribution of the different species of *Tarsius* (redrawn and adapted from Hill, 1955)

2.1.1.1. *CALLITHRIX ARGENTATA* (Linnaeus 1771)

Reproduced by courtesy of W. C. Osman Hill

2.1.1.1. *CALLITHRIX ARGENTATA* (Linnaeus 1771)

GEOG DIST: South America BDY WT: 175–360g♂; 167–335g♀
HD/BDY LTH: 175–220mm♂; 160–240mm♀ TAIL LTH: 245–375 mm♂; 250–385 mm♀
COAT COL: silvery white or light brown; tail black

SYNONYMS

Simia argentatus Linnaeus 1771
Jacchus melanurus Geoffroy 1812
Mico argentatus Lesson 1840
Jacchus leucomerus Gray 1845

VERNACULAR NAMES

Silberäffchen
Black-tailed marmoset
Ouistitì mélanure
Silvery marmoset
Mico
Sahuim branco
Macaquinho branco
Chauim branco
Titi plateado
Schwarzschwänziges

2.1.1.2.
CALLITHRIX HUMERALIFER (E. Geoffroy 1812)

Reproduced by courtesy of Gabrielle Buckley

GEOG DIST: South America   BDY WT: 175–360g♂; 170–335g♀
HD/BDY LTH: 175–220mm♂; 160–240mm♀ TAIL LTH: 245–375 mm♂; 250–385 mm♀
COAT COL: brownish-black; face segmented with pink muzzle; tail black at base

SYNONYMS

Jacchus humeralifer E. Geoffroy 1812
Hapale humeralifer Kuhl 1820
Callithrix chrysoleuca Wagner 1842
Hapaie chrysoleucos Wagner 1842
Mico sericeus Gray 1868
Micoella sericeus Gray 1870
Micoella chrysoleucos Gray 1870
Mico leucippe Thomas 1922

2.1.1.2.
CALLITHRIX HUMERALIFER (E. Geoffroy 1812)

White-shouldered marmoset
Ouistiti à camail
Golden marmoset
Silky marmoset
Yellow-legged marmoset
Sahuim branco
Titi sedosa
Titi de piernas doradas
Mico bianca
Gelbschwänziges
Silberäffchen
Santarem marmoset

CALLITHRIX JACCHUS (Linnaeus 1766)

Reproduced by courtesy of Zoologischer Garten, Hannover

GEOG DIST: South America   BDY WT: 175–360g♂; 100–335g♀
HD/BDY LTH: 175–220mm♂; 160–240mm♀   TAIL LTH: 245–375 mm♂; 250–385 mm♀
COAT COL: black and grey or brown; tail ringed black and grey

2.1.1.3. *CALLITHRIX JACCHUS* (Linnaeus 1766)

Hapale jacchus Linnaeus 1758
Simia jacchus Linnaeus 1758
Hapale penicillata Wagner 1810
Simia aurita Humboldt 1811
Simia geoffroyi Humboldt 1811
Callithrix pennicillata E. Geoffroy 1812
Callithrix geoffroy Humboldt 1812
Hapale leucocephala Geoffroy 1812
Jacchus aurita Geoffroy 1812
Jacchus leucocephalus Geoffroy 1812
Jacchus penicillatus Geoffroy 1812
Jacchus vulgaris Geoffroy 1812
Hapale albifrons Desmarest 1820
Hapale aurita Kuhl 1820
Hapale geoffroyi Kuhl 1820
Hapale leucotis Lesson 1840
Jacchus maximiliani Reichenbach 1862
Jacchus trigonifer Reichenbach 1862
Callithrix flaviceps Thomas 1903
Callithrix leucocephala Elliot 1913
Hapale coelestis Ribeiro 1924
Callithrix aurita Vieira 1955

VERNACULAR NAMES

Weisspinseläffchen
Saguim
Common marmoset
Commun ouistiti
Titi comùn
Buff-headed marmoset
Titi de cabeza rubia
White-eared marmoset
Titi de pinceles blancos
Ouistiti oreillard
Schwarzpinseläffchen

Black-eared marmoset
Ouistiti à pince au(x) moir(s)
Black-pencilled marmoset
Titi de pinceles negros
Pinzelaffe
Geoffroy's marmoset
White-fronted marmoset
Titi caratinga
Titi de cabeza blanca
Ouistiti à tête blanche

2.1.2.1.   *CEBUELLA PYGMAEA*   (Spix 1823)

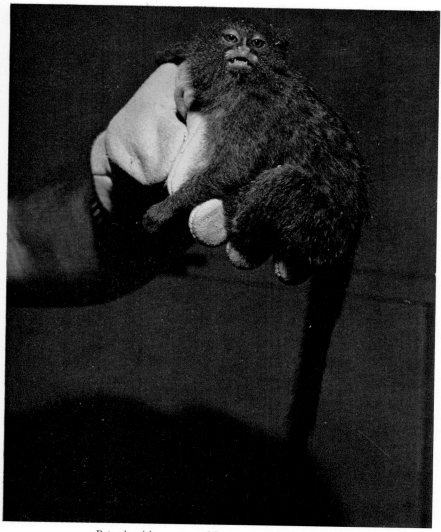

Reproduced by courtesy of Department of Medical Illustration, O.R.P.R.C.

GEOG DIST: South America   BDY WT: *ca* 150g
HD/BDY LTH: 130–145 mm   TAIL LTH: 200–210 mm
COAT COL: brown; yellow/green subterminal colouration of hairs giving tawny
   grizzled effect on back, neck, flanks; underpart white or fawn

2.1.2.1. *CEBUELLA PYGMAEA* (Spix 1823)

SYNONYMS

Jacchus pygmaeus Spix 1823
Hapale pygmaea I. Geoffroy 1851
Callithrix pygmaea Elliot 1913

VERNACULAR NAMES

Zwergseidenäffchen
Pygmy marmoset
Ouistiti mignon
Chichico
Leoncito
Saguim leãosinho
Titi enano

LEONTIDEUS ROSALIA   (Linnaeus 1766)

Reproduced by courtesy of Bernhard Grzimek

GEOG DIST: S.E. Brazil
HD/BDY LTH: 230–370 mm
COAT COL: uniformly red-gold

BDY WT: 553♂; 480♀
TAIL LTH: 300–360 mm

111

2.1.3.1.   *LEONTIDEUS ROSALIA*   (Linnaeus 1766)

SYNONYMS

Simia rosalia Linnaeus 1766
Simia leonina Humboldt 1811
Midas rosalia E. Geoffroy 1812
Jacchus albifrons Desmarest 1820
Midas leoninus Bates 1863
Hapale rosalia Schlegel 1876
Jacchus rosalia Desmarest 1820
Leontopithecus marikina Lesson 1840
Marikina albifrons Reichenbach 1862
Marikina rosalia Reichenbach 1862
Leontopithecus rosalia Gray 1870

VERNACULAR NAMES

Löwenäffchen
Lion marmoset
Tamarin rosalia
Petit singe-lion
Silky tamarin
Golden marmoset
Sahuim piranga
Mico leão vermelho
Titi dorado
Leontocito de mocoa
Marikina aurore

2.1.3.2. *LEONTIDEUS CHRYSOMELAS* (Kuhl 1820)

*Reproduced by courtesy of Adelmar F. Coimbra-Filho
and Russell A. Mittermeir*

GEOG DIST: South America
HD/BDY LTH: 230–370 mm
COAT COL: black, gold forehead and inner side of limbs

BDY WT:
TAIL LTH: 300–360 mm

SYNONYMS

Midas chrysomelas Kuhl 1820
Jacchus chrysomelas Desmarest 1820
Hapale chrysomelas Wied 1823
Leontocebus ater B. Lesson 1840
Marikina chrysomelas Reichenbach 1862
Leontopithecus chrysomelas Gray 1870

VERNACULAR NAMES

Golden-headed tamarin
Black-and-gold tamarin
Wied's tamarin
Sahuim una
Titi negro y oro

113

2.1.3.3. *LEONTIDEUS CHRYSOPYGUS* (Mikan 1823)

Geog Dist: South America   Bdy Wt:
Hd/Bdy Lth: 230–370 mm   Tail Lth: 300–360 mm
Coat Col: black; forehead and inside of limbs gold

SYNONYMS

Jacchus chrysopygus Mikan 1820
Hapale chrysopyga Wagner 1840
Marikina chrysopygus Reichenbach 1862
Midas chrysopygus Pelzeln 1883
Midas (Tamarinus) chrysopygus Trouessaert 1904
Leontocebus (Tamarinus) chrysopygus Elliot 1913
Mystax chrysopygus Thomas 1922

VERNACULAR NAMES

Golden-rumped tamarin
Yellow-tailed tamarin
Sagüi
Sauim
Marikina à queue jaune

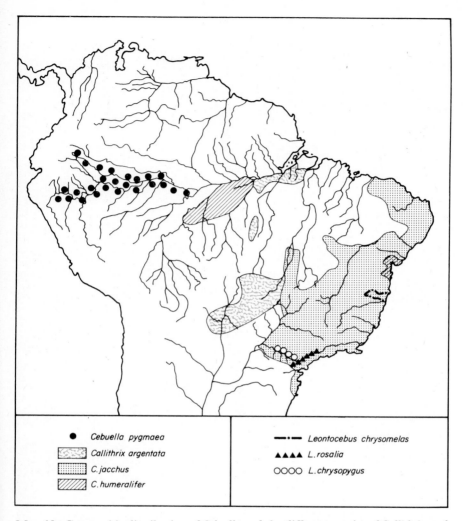

● Cebuella pygmaea	—·— Leontocebus chrysomelas
Callithrix argentata	▲▲▲▲ L.rosalia
C.jacchus	OOOO L.chrysopygus
C.humeralifer	

MAP 12. Geographic distribution of *Cebuella* and the different species of *Callithrix* and *Leontocebus* (redrawn and adapted from Hill, 1957 and Napier, 1967)

SAGUINUS TAMARIN (Link 1795)

Reproduced by courtesy of Tierbilder Okapia

GEOG DIST: South America   BDT WT. 365–340g♂; 265–395♀
HD/BDY LTH: 170–310mm♂; 155–280mm♀ TAIL LTH: 275–420 mm♂; 325–425 mm♀
COAT COL: black with rust marbling on back

2.1.4.1.   *SAGUINUS TAMARIN*   (Link 1795)

Saguinus ursulus Hoffmannsegg 1807
Midas ursulus Geoffroy 1812
Hapale ursula Wagner 1840
Jacchus tamarin Wallace 1854
Cercopithecus ursulus Elliot 1911
Tamarin tamarin

VERNACULAR NAMES

Mohrenäffchen
Negro tamarin
Tamarin nègre
Saguim preto
Titi negro

2.1.4.2. *SAGUINUS MIDAS* (Linnaeus 1758)

GEOG DIST: South America BDY WT: 265–340g♂; 265–395♀
HD/BDY LTH: 170–310mm♂; 155–280mm♀ TAIL LTH: 275–420 mm♂; 325–425 mm♀
COAT COL: black, rust marbling on back; orange or yellow hands and feet

SYNONYMS

Simia midas Linnaeus 1758
Simia lacépèdii Fischer 1806
Midas rufimanus Geoffroy 1812
Midas ursulus Gray 1870
Hapale midas Schlegel 1876

Midas midas Forbes 1894
Leontopithecus midas Thomas 1911
Cercopithecus midas Elliot 1913
Mistax midas Pocock 1917
Tamarin midas Lima 1945

VERNACULAR NAMES

Rufous-handed tamarin
Tamarin aux mains rousses
Rothandäffchen
Red-handed tamarin

Lacépède's tamarin
Saguim de mão ruiva
Titi de manos rubias

Reproduced by courtesy of Tierbilder Okapia

GEOG DIST: South America   BDY WT: 265–340g♂; 265–395g♀
HD/BDY LTH: 170–310mm♂; 155–280mm♀ TAIL LTH: 275–420 mm♂; 325–425♀
COAT COL: whitish or yellow/white; 3-zoned pattern on dorsum; face, ears, hands, feet black

2.1.4.3.   *SAGUINUS FUSCICOLLIS*   (Spix 1823)

Hapale illigeri Pucheran 1845
Midas illigeri Geoffroy 1851; Forbes 1894; Goeldi 1907
Leontocebus illigeri Elliot 1913
Leontecebus mounseyi Thomas 1920
Midas tripartitus Milne-Edwards 1878
Mystax bluntschlii Matschie 1915
Midas lagonotus Espada 1870
Tamarinus illigeri Pucheran 1845
Saguinus illigeri Pucheran 1845
Midas flavifrons I. Geoffroy and Deville 1848
Hapale nigrifrons I. Geoffroy 1850
Midas nigrifrons pebilis Thomas 1928
Leontecebus nigrifrons Geoffroy 1850
Saguinus fuscus Lesson 1840
Saguinus weddelli Deville 1849
Saguinus lagonotus Espada 1870
Saguinus devilli I. Geoffroy 1850
Midas leucogenys Gray 1866
Leontocebus pacator Thomas 1914
Leontocebus purillus Thomas 1914
Saguinus melanoleucus Ribeiro 1912
Mico melanoleucus Miranda Ribeiro 1912
Mystax melanoleucus Thomas 1922
Leontocebus hololeucus Pinto 1937
Mystax imberbis Lönnberg 1940
Callithrix melanoleuca Lima 1944
Tamarin melanoleucus Serra 1950

VERNACULAR NAMES

Red-mantled tamarin
Illiger's tamarin
Golden-mantled tamarin
Chichico
Brown-headed tamarin
Tamarino de cabeza amarilla

Weddell's tamarin
Monito leoncito
Weisslippenäffchen
White tamarin
Saguim blanco
Titi de cara negra

2.1.4.4. *SAGUINUS GRAELLSI* (Espada 1870)

Reproduced by courtesy of W. C. Osman Hill

2.1.4.4.   *SAGUINUS GRAELLSI*   (Espada 1870)

GEOG DIST: South America   BDY WT: 265–340g$\male$; 265–395$\female$
HD/BDY LTH: 170–310mm$\male$; 155–280mm$\female$ TAIL LTH: 275–420 mm$\male$; 325–425 mm$\female$
COAT COL: mantle agouti or melano-agouti; 2-zonal pattern on dorsum; face black; lower back, thighs, underparts olivaceous or buff/brown

SYNONYMS

Midas graellsi Espada 1870
Tamarinus graellsi

VERNACULAR NAMES

Rio Napo tamarin
Yurumuruchi
Uxpachichico

2.1.4.5.   *SAGUINUS NIGRICOLLIS*   (Spix 1823)

GEOG DIST: South America   BDY WT: 265–340 g♂; 265–395g♀
HD/BDY LTH: 170–310mm♂; 155–280mm♀ TAIL LTH: 275–420 mm♂; 325–425 mm♀
COAT COL: mantle black; 2-zonal pattern on dorsum; face black; lower back, rump,
   thighs reddish or mahogany, under parts reddish mixed with black

SYNONYMS

Midas nigricollis Spix 1823
Midas rufoniger I. Geoffroy and Deville 1848
Leontocebus nigricollis Elliot 1913
Leontocebus purillus Thomas 1914
Tamarinus nigricollis

VERNACULAR NAMES

Black and red tamarin
Sagui

2.1.4.6. *SAGUINUS IMPERATOR* (Goeldi 1907)

GEOG DIST: South America   BDY WT: 265–340g♂; 265–395g♀
HD/BDY LTH: 170–310mm♂; 155–280mm♀ TAIL LTH: 275–420 mm♂; 325–425 mm♀
COAT COL: grey with reddish tail; long, drooping, white moustache

2.1.4.6. *SAGUINUS IMPERATOR* (Goeldi 1907)

SYNONYMS

Midas imperator Goeldi 1907
Leontocebus imperator Elliot 1913
Marikina imperator Hershkovitz 1949
Tamarinus imperator

VERNACULAR NAMES

Emperator tamarin
Emperor tamarin
Tamarino imperial
Sagui de bigode
Kaiser tamarin

2.1.4.7.   *SAGUINUS MYSTAX*   (Spix 1823)

Reproduced by courtesy of W. C. Osman Hill

GEOG DIST: South America   BDY WT: 265–340g♂; 265–395g♀
HD/BDY LTH: 170–310mm♂; 155–280mm♀   TAIL LTH: 275–420 mm♂; 325–425 mm♀
COAT COL: mainly black with conspicuous white moustache

SYNONYMS

Midas mystax Spix 1823; I. Geoffroy 1851; Gray 1870
Leontocebus mystax Elliot 1913; Cabrera 1917
Mystax mystax Thomas 1922; Lönnberg 1940
Tamarin mystax Lima 1945
Marikina mystax Hershkovitz 1949
Tamarinus mistax
Saguinus pileatus I. Geoffroy and Deville 1848
Midas pileatus I. Geoffroy and Deville 1848; Goeldi 1907
Leontocebus pileatus Elliot 1913
Saguinus pluto Lönnberg 1926
Mystax pluto Lönnberg 1926
Tamarin pluto Lima 1945

VERNACULAR NAMES

Schnurrbartäffchen
Moustached tamarin
Saguim preto de bigode branco
Saguim de boca branca
Tamarin à moustaches
Weissgesichtäffchen
Red-capped tamarin

Bonneted tamarin
Brown-headed tamarin
Tamarin à calotte rousse
Tamarino de cabeza roja
Lönnberg's tamarin
Chauim

126

2.1.4.8. *SAGUINUS LABIATUS* (E. Geoffroy 1812)

GEOG DIST: South America
HD/BDY LTH: 170– 310 mm♂
BDY WT: 265–340g♂; 265–395g♀
TAIL LTH: 275–420 mm♂; 325–425 mm♀
COAT COL: blackish brown back; black face with white around lips; black limbs and tail; under parts and inside of limbs orange-red

SYNONYMS

Simia labiata Humboldt 1812
Midas labiatus Geoffroy 1812
Leontocebus labiatus Geoffroy 1812
Jacchus rufiventer Gray 1843
Midas elegantulus Slack 1862
Midas erythrogaster Reichenbach 1862
Midas thomasi Goeldi 1907
Midas griseovertex Goeldi 1907

VERNACULAR NAMES

Red-bellied, white-lipped tamarin
Saguim de bigode branco
Tamarino labiado
Weisslippenäffchen
Rotbauchäffchen
Tamarin labié

127

2.1.4.9. *SAGUINUS GEOFFROYI*   (Pucheran 1845)

Reproduced by courtesy of New York Zoo

GEOG DIST: East of Panama zone   BDY WT: 300–510g♂; 335–565g♀
HD/BDY LTH: 220–245mm♂; 250mm♀   TAIL LTH: 360–380 mm♂; 370 mm♀
COAT COL: dark brown; white under parts and limbs

SYNONYMS

Hapale geoffroyi Pucheran 1845   *Midas geoffroyi* Geoffroy 1845
Jacchus spixii (Reichenbach 1862)   *Oedipomidas geoffroyi* Reichenbach 1862
Oedipomidas salaquiensis Elliot 1912 *Oedipus geoffroyi* Gray 1870
Midas oedipus Spix 1823   *Leontocebus geoffroyi* Anthony 1916
Oedipus titi Lesson 1840   *Oedipomidas spixi*

VERNACULAR NAMES

Geoffroy's tamarin
Titi
Bichichi
Perückenäffchen

128

2.1.4.10. *SAGUINUS OEDIPUS* (Linnaeus 1758)

GEOG DIST: North Colombia

HD/BDY LTH: 220–245 mm♂; 250 mm♀

COAT COL: dark brown back; white underparts and limbs

BDY WT: 300–510g♂; 335–565g♀

TAIL LTH: 360–380 mm♂; 370 mm♀

SYNONYMS

Simia oedipus Linnaeus 1758

Seniocebus meticulosus Elliot 1912

Midas oedipus Humboldt 1805–1812

Callithrix sciurea Schott 1861

Oedipomidas oedipus Reichenbach 1862

Leontopithecus oedipus Thomas 1911

Marikina oedipus Hershkovitz 1949

VERNACULAR NAMES

Pinche marmoset

Tamarin pinche

Cotton-head tamarin

Titi

Bichichi

Perückenäffchen

Cotton-tops

Liszt monkey

2.1.4.11.   *SAGUINUS LEUCOPUS*   (Günther 1876)

Drawing by Kostas, Turin

GEOG DIST: South America
BDY WT:
HD/BDY LTH: 220–290 mm♂; 210–250 mm♀   TAIL LTH: 350–370 mm♂; 235–420 mm♀
COAT COL: brown with yellow or rust red underparts; forearms and hands whitish;
long silvery hair on cheeks and forehead

SYNONYMS

Hapale leucopus Günther 1876
Callithrix leucopus Elliot 1913
Seniocebus pegasis Elliot 1913
Oedipomidas leucopus Elliot 1914
Mystax leucopus Cabrera and Yepes 1940
Marikina leucopus Hershkovitz 1949
Tamarinus leucopus

VERNACULAR NAMES

White-footed tamarin
Tamarino de manos blancas
Titi

130

2.1.4.12. *SAGUINUS INUSTUS* (Schwarz 1951)

GEOG DIST: South America   BDY WT:
HD/BDY LTH: 220–290 mm♂; 210–250 mm♀ TAIL LTH: 350–370 mm♂; 235–420 mm♀
COAT COL: black, brown back and flanks; face bare, black

SYNONYMS

Leontocebus midas Schwarz 1951
Tamarin inustus

VERNACULAR NAMES

2.1.4.13.   *SAGUINUS BICOLOR*   (Spix 1823)

Reproduced by courtesy of Tierbilder Okapia

GEOG DIST: South America   BDY WT:
HD/BDY LTH: 220–290 mm♂; 210–250 mm♀   TAIL LTH: 350–370 mm♂; 235–420 mm♀
COAT COL: chest, forelimbs white; hind parts yellow-brown; face bare, black

132

2.1.4.13.   *SAGUINUS BICOLOR*   (Spix 1823)

SYNONYMS

Midas bicolor Spix 1823
Hapale bicolor Wagner 1855
Seniocebus bicolor Gray 1870
Tamarin (Oedipomidas) bicolor Tate 1919
Marikina bicolor

VERNACULAR NAMES

Pied tamarin
Zweifarbentamarin
Saguim de duas colores
Tamarino bicolor
Tamarin bicolore
Manteläffchen

2.1.4.14. *SAGUINUS MARTINSI* (Thomas 1912)

GEOG DIST: South America BDY WT:
HD/BDY LTH: 220–290 mm♂; 210–250 mm♀ TAIL LTH: 350–370 mm♂; 235–420 mm♀
COAT COL: brown; yellowish forearms and hands; face bare, black

SYNONYMS

Leontocebus martinsi Thomas 1912
Seniocebus martinsi Elliot 1913
Oedipomidas martinsi Elliot 1914
Tamarin martinsi Tate 1939
Marikina martinsi

VERNACULAR NAMES

Martin's tamarin
Sagui
Saim
Soim

135

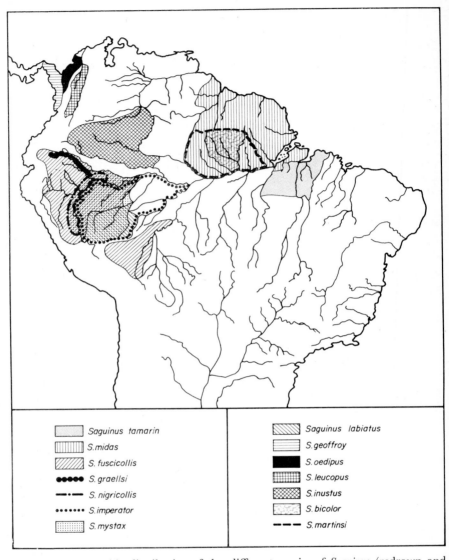

	Saguinus tamarin		Saguinus labiatus
	S.midas		S.geoffroy
	S. fuscicollis		S.oedipus
	S. graellsi		S.leucopus
	S. nigricollis		S.inustus
	S.imperator		S. bicolor
	S.mystax		S.martinsi

MAP 13. Geographic distribution of the different species of *Saguinus* (redrawn and adapted from Hill, 1957 and Hershkovitz, 1966)

CALLIMICO GOELDII (Thomas 1904)

Reproduced by courtesy of Rühmekorf, Hannover

GEOG DIST: South America
HD/BDY LTH: 190–215 mm♂; 190 mm♀
COAT COL: predominantly black

BDY WT:
TAIL LTH: 255–325 mm♂; 270 mm♀

2.2.1.1. *CALLIMICO GOELDII* (Thomas 1904)

Midas goeldi Thomas 1904
Callimico snethlageri Ribeiro 1912
Callimidas snethlageri Ribeiro 1912
Callithrix goeldi Elliot 1913

VERNACULAR NAMES

Springtamarin
Goeldi's marmoset
Tamarin de Goeldi
Goeldi's monkey
Goeldi's marmoset

2.3.1.1.   *AOTES TRIVIRGATUS*   Humboldt 1811

Reproduced by courtesy of Bernhard Grzimek

GEOG DIST: Central and South America   BDY WT: 825–1020 mm♂; 780–1250 mm♀
HD/BDY LTH: 240–475 mm   TAIL LTH: 220–420 mm
COAT COL: brown or grey, lighter ventrum; 3 longitudinal dark streaks on head

2.3.1.1. *AOTES TRIVIRGATUS* Humboldt 1811

SYNONYMS

Pithecia miriquouina Geoffroy 1812
Callithrix infulatus v. Hasselt and Kuhl 1820
Nyctipithecus vociferans Spix 1823
Nyctipithecus felinus Spix 1823
Nyctipithecus lemurinus Geoffroy 1843
Aotes microdon Dollman 1909
Aotes griseimembra Elliot 1912
Pithecia hirsuta Schott 1861
Aotes lemurinus Allen 1904
Aotes zonalis Goldman 1914
Aotes bipunctatus Bole 1937
Aotes rufipes Sclater 1872
Aotes vociferans Spix 1823
Nyctipithecus commersonii Gray 1873
Nyctipithecus villosus Gray 1847
Nyctipithecus hirsutus Gray 1870
Aotes lanius Dollman 1909
Aotes aversus Elliot 1913
Aotes pervigilis Elliot 1913
Aotes humboldtii Illiger 1811
Aotes duruculi Lesson 1840
Aotes oseryi Geoffroy and Deville 1848
Aotes spixii Pucheran 1857
Aotes senex Dollman 1909
Aotes gularis Dollman 1909
Cheirogaleus commersonii Vigors and Horsfield 1828
Nyctipithecus azarae Ihering, Goeldi and Hagmann 1906
Aotes nigriceps Dollman 1909
Aotes boliviensis Elliot 1907
Aotes miconax Thomas 1907
Aotes bidentatus Lönnberg 1941
Aotes roberti Dollman 1909
Aotus azarae Thomas 1903
Simia (Pithecia) azarae Humboldt 1812
Aotes miriquina Desmarest 1820
Nyctipithecus trivirgatus Rengger 1830
Aotes azarae Humboldt 1812

2.3.1.1. *AOTES TRIVIRGATUS* Humboldt 1811

Nachtäffen
Owl-faced monkey
Douroucouli
Night monkey
Devil monkey
Ei-A
Mico de noche
Singe de nuit
Cusi-cusi
Mico dormilon
Mono de noche de Tolima
Ya
Cara rayada

Tutacusillo
Euh-à
Lärmender Nachtaffe
Macaco da noite
Miriquoinà
Katzen Nachtaffe
Mono de noche andino
Mono de noche de cabeza negra
Mono nocturno
Miriquinà
Macaco de nocte
Adufeiro macaco
Mirikinà

Reproduced by courtesy of San Diego Zoo

GEOG DIST: South America   BDT WT:
HD/BDY LTH: TAIL LTH: > Hd/Bdy Lth
COAT COL: dark brown; long buff-tipped hairs on back; tail red-brown

2.3.2.1.   *CALLICEBUS CUPREUS*   (Spix 1823)

SYNONYMS

Callithrix discolor Geoffroy and Deville 1848
Callicebus ustofuscus Elliot 1907
Callicebus calligatus (Wagner 1842)
Callicebus brunneus (Wagner 1842)
Callithrix ornatus (Gray 1866)
Callicebus moloch (Hoffmannsegg 1807)
Callicebus hoffmannsi Thomas 1908
Callicebus remulus Thomas 1908
Callicebus emiliae Thomas 1911
Callicebus cinerascens (Spix 1823)
Callithrix cuprea Spix 1823
Callicebus personnatus (E. Geoffroy 1812)
Cebus moloch Hoffmannsegg 1807
Callithrix moloch Forbes 1894
Callicebus melanops Vigors 1829
Callicebus ollalae Lönnberg 1939

VERNACULAR NAMES

Springaffe
Titi monkeys
Red titi
Masked titi
Japussà
Uapo
Oyapuca
Sahui rojo
Callicèbe roux
Roter Springaffe
Kupferfarbiger Springaffe
Ouappo
Zocayo
Tzocallo
Callicèbe discolore

Cuivé
Ashy titi
Saua
Otôhò
Sahui ceniciento
Sagoin à couleur de souris
Ascfarbiger Springaffe
Orabassu titi
Arabassù
Sahui gauzu
Sagoin à tête et mains noires
Callicèbe à masque
Schwarzköpfiger Springaffe
Ollala's titi

2.3.2.2. *CALLICEBUS TORQUATUS* (Hoffmannsegg 1807)

GEOG DIST: South America
HD/BDY LTH: 310–375 mm
COAT COL: reddish black; tail black

BDY WT:
TAIL LTH: 420–495 mm

2.3.2.2. *CALLICEBUS TORQUATUS* (Hoffmannsegg 1807)

SYNONYMS

Callithrix torquata Hoffmannsegg 1807
Simia lugens Humboldt 1811
Saguinus vidua Lesson 1840
Callicebus lucifer Thomas 1914

VERNACULAR NAMES

Collared titi
Yellow-handed titi
White-Collared titi
Widow monkey
Witwenaffe
Japussa
Viudita
Sahui de collar
Callicèbe à fraise
Callicèbe à collier

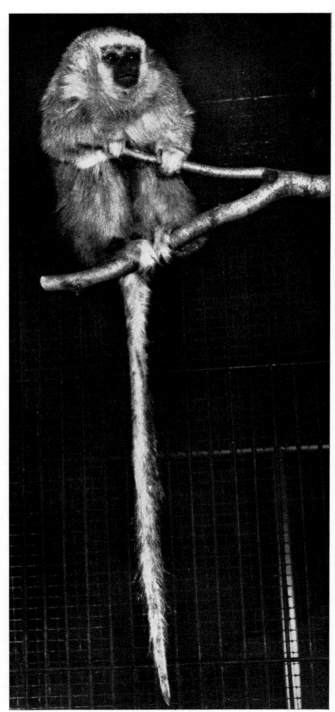

Reproduced by courtesy of W. C. Osman Hill

2.3.2.3. *CALLICEBUS GIGOT* (Spix 1823)

GEOG DIST: South America
HD/BDY LTH: 290–390 mm
BDY WT: 681g♂
TAIL LTH: 330–480 mm
COAT COL: grey, reddish or brown; tail dark grey, sometimes with whitish tip

SYNONYMS

Callithrix gigot Spix 1823
Callithrix gigo Gray 1870

VERNACULAR NAMES

Grey titi
Sahui gris
Guigo
Chigot
Chico
Gigot Springaffe

2.3.3.1. *PITHECIA PITHECIA* (Linnaeus 1766)

Reproduced by courtesy of San Diego Zoo

GEOG DIST: South America   BDY WT: 1580g♂; 1410g♀
HD/BDY LTH: 355–480 mm♂; 300–425 mm♀ TAIL LTH: 315–510 mm♂; 255–545 mm♀
COAT COL: predominantly black/dark grey; ♂black muzzle surrounded by creamy white; ♀light-coloured oblique paranasal streaks

SYNONYMS

Simia pithecia Linnaeus 1766   *Pithecia adusta* Illiger 1815
Simia leucocephala Audebert 1772 *Pithecia capillamentosa* Spix 1823
Pithecia rufiventer Humboldt 1812 *Pithecia pogonias* Gray 1842
Pithecia nocturna Illiger 1815   *Pithecia chrysocephala* Geoffroy 1850

VERNACULAR NAMES

Aliki
Wanaka
Huruwe
Sakiwinki
Weisskopfaffe
White-headed saki
Golden headed saki
Saki à tête blanche
Saki à tête dorée
Pale-headed saki
Waiti-Feici-Wanaku
Black-Feici-Wanaku

Wanaku
Fox-tailed monkey
Bisa monkey
Beesa monkey
Waiti-Facei
Blaki-Facei-Wanaku
Sacca winki
Wanakoe
Weisskopfsaki
Yarké
Yarqué
Thari-ki

2.3.3.2. *PITHECIA MONACHUS*   E. Geoffroy 1812

Reproduced by courtesy of Tierbilder Okapia

GEOG DIST: South America   BDT WT: 1580g♂; 1410g♀
HD/BDY LTH: 355–480 mm♂; 300–425 mm♀ TAIL LTH: 315–510 mm♂; 255–545 mm♀
COAT COL: dark grey/brindled; face dark, pale oblique stripes either side of nose

2.3.3.2. *PITHECIA MONACHUS* E. Geoffroy 1812

SYNONYMS

Pithecia hirsuta Spix 1823
Pithecia inusta Spix 1823
Pithecia capillamentosa Spix 1823
Pithecia quapa Pöppig 1835–36
Pithecia irrorata Gray 1840 (1844)
Pithecia albicans Gray 1860
Pithecia milleri Allen 1914
Pithecia napensis Lönnberg 1938

VERNACULAR NAMES

Zottelaffe
Mönchsaffe
Hairy saki
Monksaki
Saki moine
Whitish saki
Parauacu
Macaco cabeludo
Zottiger Bärenaffe
Saki à perruque
Paraguaçu

2.3.4.1. *CHIROPOTES SATANAS* (Hoffmannsegg 1807)

2.3.4.1. *CHIROPOTES SATANAS* (Hoffmannsegg 1807)

GEOG DIST: South America
BDY WT: 2770–3130g♂
HD/BDY LTH: 400–410mm♂; 460mm♀
TAIL LTH: 380 mm♂; 350 mm♀
COAT COL: black with chestnut brown colouration on back

SYNONYMS

Chiropotes couxio Lesson 1840
Chiropotes ater Gray 1870
Cebus satanas Hoffmannsegg 1807
Simia satanas Humboldt 1811
Pithecia satanas Wagner 1855
Chiropotes chiropotes Humboldt 1811
Simia chiropotes Humboldt 1811
Brachyurus israelitica Spix 1823
Simia sagulata Traill 1821
Pithecia chiropotes Forbes 1894

VERNACULAR NAMES

Satansaffe
Black saki
Satansaap
Saki noir
Saki satanique
Cuxiù negro
Judenaffe

Red-backed saki
Saki capucin
Jacketed monkey
Schweifaffe
Kwatta swageri
Coxiù comun
Mono capichino judeù

2.3.4.2. *CHIROPOTES ALBINASA* (I. Geoffroy and Deville 1848)

Reproduced by courtesy of W. C. Osman Hill

GEOG DIST: South of Amazon River
HD/BDY LTH: 400–410 mm♂; 460 mm♀
COAT COL: black with pinkish nose and upper lip

BDY WT: 2770–3130g♂
TAIL LTH: 380 mm♂; 350 mm♀

SYNONYMS

Pithecia albinasa I. Geoffroy and Deville 1848

VERNACULAR NAMES

Weissnasenaffe
White-nosed saki
Saki à nez blanc
Piroculù
Guxiù de nariz branco

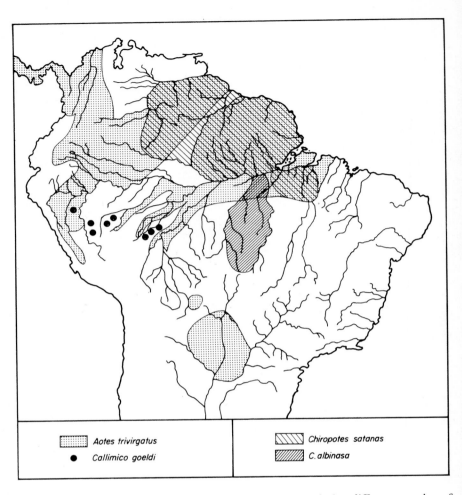

MAP 14. Geographic distribution of *Aotes*, *Callimico* and the different species of *Chiropotes* (redrawn and adapted from Hill, 1957–60)

The legend of the map reads:

Aotes trivirgatus	*Chiropotes satanas*
● *Callimico goeldi*	*C. albinasa*

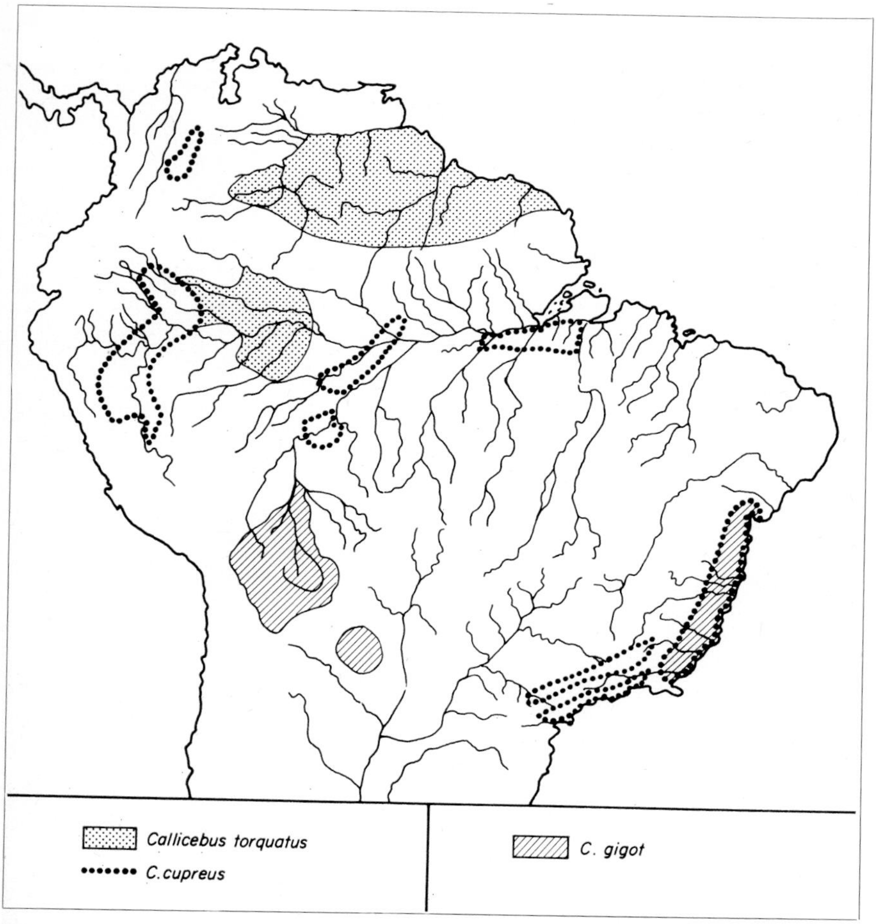
Callicebus torquatus
C. cupreus
C. gigot

MAP 15. Geographic distribution of the different species of *Callicebus* (redrawn and adapted from Hill, 1960)

Reproduced by courtesy of Tierbilder Okapia

2.3.5.1.   *CACAJAO CALVUS*   (I. Geoffroy 1845)

GEOG DIST: Brazil: between Japura
and Ica Rivers   BDY WT:
HD/BDY LTH: 435–485 mm♂; 365–445 mm♀ TAIL LTH: 155–185 mm♂; 150–165 mm♀
COAT COL: Silver grey or white; face and forehead pink, bare

SYNONYMS

Brachyurus calvus I. Geoffroy 1847
Ouakaria calva Gray 1870
Pithecia alba Schlehel 1876
Cacajao roosevelti J. A. Allen 1914

VERNACULAR NAMES

Scharlachgesicht
White uakari
Scarlet-faced monkey
Ouakari chauve
Bald uakari
Red-faced uakari
Scarlet-fever faced monkey
Ouajari chauve
Black uakari
Uacari blanco
Macari de cara blanca
Black makari

157

2.3.5.2. *CACAJAO RUBICUNDUS* (I. Geoffroy and Deville 1848)

Reproduced by courtesy of W. C. Osman

GEOG DIST: South America   BDY WT:
HD/BDY LTH: 435–485 mm♂; 365–445 mm♀ TAIL LTH: 155–185 mm♂; 150–165 mm♀
COAT COL: red-brown; face and forehead crimson, bare

SYNONYMS

Brachyurus rubicundus I. Geoffroy and Deville 1848
Ouakaria rubicunda Gray 1870
Pithecia rubicunda Schlegel 1876

VERNACULAR NAMES

Roter Uakari
Red uakari
Ouakari rubicond
Uacari vermello
Macaco inglés
Macaco acarý
Uacari rojo

2.3.5.3. *CACAJAO MELANOCEPHALUS* (Humboldt 1811)

Reproduced by courtesy of San Diego Zoo

159

2.3.5.3. *CACAJAO MELANOCEPHALUS* (Humboldt 1811)

GEOG DIST: South America: between
  Orinoco/Negro and Japura Rivers   BDY WT:
HD/BDY LTH: 435–485 mm♂; 365–445 mm♀ TAIL LTH: 155–185 mm♂; 150–165 mm♀
COAT COL: chestnut with black extremities; face and forehead bare, black

SYNONYMS

Simia melanocephala Humboldt 1812
Pithecia melanocephala E. Geoffroy 1812
Brachyurus ouakary Spix 1823
Ouakaria spixii Gray 1849
Brachyurus melanocephalus W. A. Forbes 1880

VERNACULAR NAMES

Schwarzkopfuakari
Mono feo
Black-headed uakari
Ouakari à tête noire
Häblicher Affe
Cacajao de cabeza preta
Macaco mal acabado
Mono rabòn
Uacari de cara negra
Chucuto
Chucuzo
Cariuri
Cacahao

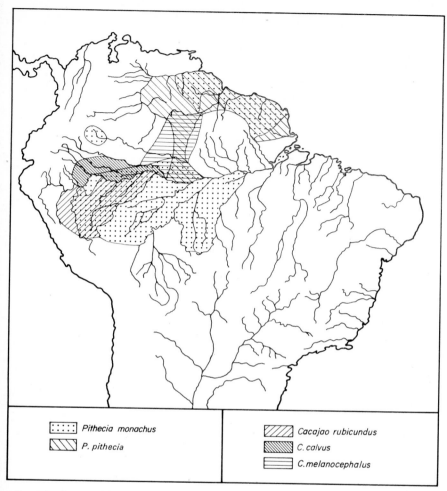

MAP 16. Geographic distribution of the different species of *Pithecia* and *Cacajao* (redrawn and adapted from Hill, 1960)

2.3.6.1.   *ALOUATTA VILLOSA*   (Gray 1845)

Drawing by Kostas, Turin

GEOG DIST: Central and South America   BDY WT:
HD/BDY LTH: 465–720 mm♂; 390–575 mm♀ TAIL LTH: 490–750 mm♂; 490–701 mm♀
COAT COL: black

SYNONYMS

Mycetes villosus Gray 1845–70
Alouatta palliata (Gray 1848)
Alouatta aequatorialis Festa 1903
Mycetes niger Thomas 1880
Mycetes palliatus Gray 1848

VERNACULAR NAMES

Guatemalan howler-monkey
Villose howler
Mantelbrüllaffe
Mantled howler
Hurleur à manteau
Guariba peludo

2.3.6.2.   *ALOUATTA FUSCA*   (E. Geoffroy 1812)

Reproduced by courtesy of W. C. Osman Hill

2.3.6.2.   *ALOUATTA FUSCA*   (E. Geoffroy 1812)

GEOG DIST: South America   BDY WT:
HD/BDY LTH: 465–720mm♂; 390–575mm♀ TAIL LTH: 490–750 mm♂; 490–710 mm♀
COAT COL: brown

SYNONYMS

Simia ursina Humboldt 1812
Simia guariba Humboldt 1812
Stentor ursinus Geoffroy 1812
Stentor fuscus Geoffroy 1812
Alouatta ursina Slack 1817
Mycetes fuscus Desmarest 1820
Mycetes ursinus Wied 1823
Mycetes seniculus B. Wagner in Schreber 1840
Mycetes bicolor Gray 1845
Alouatta ursina Forbes 1894
Alouatta guariba Cabrera and Yepes 1940

VERNACULAR NAMES

Brown howler
Bugio ruivo
Guariba peludo
Guariba de la sierra
Hurleur brun
Brauner Brüllaffe

ALOUATTA SENICULUS (Linnaeus 1766)

Reproduced by courtesy of W. C. Osman Hill

GEOG DIST: South America BDY WT:
HD/BDY LTH: 465–720 mm♂; 390–575 mm♀ TAIL LTH: 490–750 mm♂; 490–710 mm♀
COAT COL: copper red

2.3.6.3.　*ALOUATTA SENICULUS*　(Linnaeus 1766)

Simia seniculus Linnaeus 1766
Simia straminea Humboldt 1811
Simia ursina Humboldt 1811
Stentor chrysurus Geoffroy 1829
Mycetes auratus Gray 1845
Alouatta macconnelli Elliot 1910
Alouatta insulanus Elliot 1910
Alouatta jaura Elliot 1910
Cebus seniculus Erxleben 1777
Mycetes seniculus Desmarest 1820
Mycetes chrysurus Schinz 1844
Micetes laniger Gray 1845
Mycetes ursinus Wagner 1848
Stentor stramineus Orgigny 1847; Gervais 1847

VERNACULAR NAMES

Roter Brüllaffe
Red howler-monkey
Hurleur roux
Guariba vermelha
Coto
Mono colorado
Araguato
Hurleur alouate
Singe rouge
Aluate
Guariba ruina

Reproduced by courtesy of W. C. Osman Hill

GEOG DIST: South America   BDY WT:
HD/BDY LTH: 465–720 mm♂; 390–575 mm♀   TAIL LTH: 490–750 mm♂; 490–710 mm♀
COAT COL: black; reddish hands, feet, tail tip

SYNONYMS

Mycetes rufimanus Van Hasselt and Kuhl 1820
Simia belzebul Linnaeus 1766
Cebus belzebul Erxleben 1777
Simia sapajus belzebul Kerr 1782
Mycetes beelzebul Gray 1845, 1870
Mycetes discolor Spix 1823
Alouatta nigerrima Lönnberg 1941
Simia beelzebul Bechstein 1800
Cebus beelzebut Latreille 1801
Mycetes flavimanus Bates 1863

VERNACULAR NAMES

Yellow-handed howler
Rufous-handed howler-monkey
Guariba da mão ruiva
Guariba de manos rubias
Hurleur à mains rousses
Gelbfüssiger Brüllaffe
Rothändiger Brüllaffe

2.3.6.5.　　*ALOUATTA CARAJA*　　(Humboldt 1811)

Geog Dist: South America   Bdy Wt:
Hd/Bdy Lth: 465–720 mm♂; 330–575 mm♀　Tail Lth: 490–750 mm♂; 490–710 mm♀
Coat Col: adult♂black; ♀and juvenile olive-buff

2.3.6.5.   *ALOUATTA CARAJA*   (Humboldt 1811)

SYNONYMS

Simia caraja Humboldt 1812
Mycetes seniculus niger Spix 1823
Mycetes niger Kuhl 1820, Gray 1845
Mycetes caraja Desmarest 1820
Cebus caraja Fischer 1829, Ribeiro 1914
Alouatta nigra Slack 1862
Mycetes seniculus Winge 1896
Stentor niger Geoffroy 1812
Mycetes barbatus Spix 1823
Alouatta nigra Trouessart 1897

VERNACULAR NAMES

Schwarzer Brüllaffe
Black howler monkey
Hurleur noir
Bugio preto
Mono aullador
Karaya

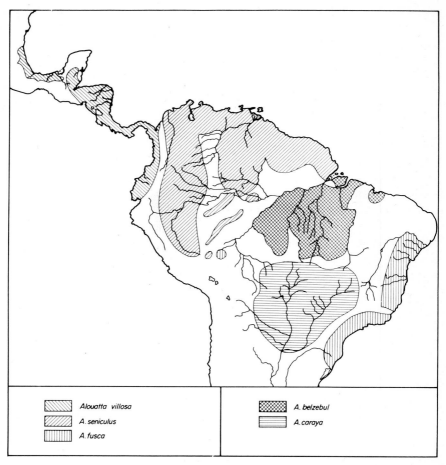

Map 17. Geographic distribution of the different species of *Alouatta* (redrawn and adapted from Hill, 1962)

2.3.7.1.   *SAIMIRI SCIUREA*   (Linnaeus 1758)

Reproduced by courtesy of Tierbilder Okapia

GEOG DIST: Central and South America   BDY WT: 550–1135g♂; 365–750g♀
HD/BDY LTH: 250–370mm♂; 225–295mm♀   TAIL LTH: 370–465 mm♂; 370–445 mm♀
COAT COL: grey-green to olive; underside and limbs white or yellow-orange; crown
   hair olive green

SYNONYMS

Callithrix sciureus E. Geoffroy 1812
Callithrix ustus I. Geoffroy 1844
Simia sciurea Humboldt 1811
Saimiris ustus Geoffroy 1844
Pithesciurus saimiri Lesson 1840
Chrysothrix nigrivittata Wagner 1846
Saimiri macrodon Elliot 1907
Saimiri boliviensis (D'Orbigny 1834)
Callithrix entomophagus D'Orbigny 1836
Saimiri madeirae Thomas 1908
Simia morta Linnaeus 1758
Chrysothrix sciurea Kaup 1835
Chrisothrix ustus Gray 1870

171

2.3.7.1.   *SAIMIRI SCIUREA*   (Linnaeus 1758)

Totenköpfchen
Common titi monkey
Squirrel monkey
Saimiri comun
Common squirrel monkey
Macaco de cheiro
Boca preta
Cai-pussu
Sapajou jaune
Saimiri aurore

Saimiri sciurin
Barizo
Frailecilo
Jurupari
Saimiri
Golden-backed squirrel-monkey
Geoffroy's squirrel-monkey
Short-tailed squirrel-monkey
Saimiri tostado
Saimiri à dos brûlé

2.3.7.2. *SAIMIRI OERSTEDII* (Reinhardt 1872)

Reproduced by courtesy of San Diego Zoo

GEOG DIST: Central and South America   BDY WT: 550–1135g♂; 365–750g♀
HD/BDY LTH: 250–370mm♂; 225–295mm♀ TAIL LTH: 370–465 mm♂; 370–445 mm♀
COAT COL: grey-green to olive, underside and limbs white or yellow-orange; crown hair black

SYNONYMS

Chrysothrix oerstedii Reinhardt 1872
Saimiri oerstedi critinellus Thomas 1904

VERNACULAR NAMES

Red-backed squirrel-monkey
Gelbes Totenköpfachen
Feuer-Zwergsai
Black-headed squirrel-monkey
Saimiri de cabeza negra
Mono chichito
Saimiri à tête noire
Black-headed titi monkey

173

| | Saimiri sciurea | | S. oerstedii |

MAP 18. Geographic distribution of the two species of *Saimiri* (redrawn and adapted from Hill, 1960)

2.3.8.1.   *CEBUS APELLA*   (Linnaeus 1758)

Reproduced by courtesy of Tierbilder Okapia

GEOG DIST: Central and South America   BDY WT: 1140–3320g♂
HD/BDY LTH: 320–565mm♂; 320–480mm♀   TAIL LTH: 340–560 mm♂; 290–510 mm♀
COAT COL: dark brown

CEBUS APELLA (Linnaeus 1758)

SYNONYMS

Simia apella Linnaeus 1758
Cebus nigritus Goldfuss 1809
Cebus niger Geoffroy 1812
Cebus variegatus Geoffroy 1812
Cebus xanthosternos Wied 1820
Cebus frontatus Kuhl 1820
Cebus robustus Kuhl 1820
Cebus hypomelas Pucheran 1857
Cebus lunatus Kuhl 1820
Cebus libidinosus Spix 1823
Cebus macrocephalus Spix 1823
Cebus xanthocephalus Spix 1823

Cebus leucogenys Gray 1865
Cebus versuta Elliot 1910
Cebus elegans I. Geoffroy 1850
Cebus vellerosus I. Geoffroy 1851
Cebus crassiceps Pucheran 1857
Cebus pallidus Gray 1865
Cebus caliginosus Elliot 1910, 1913
Simia fatuellus Linnaeus 1766
Cebus cirrifer Geoffroy 1812
Cebus macrocephalus Spix 1823
Cebus azarae Renger 1830
Cebus paraguayanus Fischer 1929

VERNACULAR NAMES

Gehaubter Kapuziner
Faunaffe
Hooded capuchin
Dickkopfkapuziner
Large-headed capuchin
Gehörnter Kapuziner
Sajou noire
Brown capuchin
Tufted capuchin
Ringtail monkey

Kaai
Cai
Macaco prego
Mico
Sajou brun
Brauner Rollaffe
Capucijnaap
Schmauz
Faun

2.3.8.2.   *CEBUS CAPUCINUS*   (Linnaeus 1758)

Reproduced by courtesy of Zoologischer Garten, Zürich

GEOG DIST: Central and South America   BDY WT: 1150–3320g♂
HD/BDY LTH: 320–565 mm♂; 323–480 mm♀ TAIL LTH: 340–560 mm♂; 290–510 mm♀
COAT COL: generally black

2.3.8.2.   *CEBUS CAPUCINUS*   (Linnaeus 1758)

Simia capucina Linnaeus 1758
Cebus hypoleucus Humboldt 1812 Reichenbach; 1862
Cebus imitator Thomas 1903

VERNACULAR NAMES

Kapuziner
White-throated capuchin
Sajou capuchin
White-faced capuchin
Carita blanca
Sai à gorge blanche
Weisschulteraffe
Weiskehlkapuziner
Capucin-apan
Iuaruka
Ibarga

2.3.8.3.   *CEBUS ALBIFRONS*   (Humboldt 1811)

2.3.8.3.   *CEBUS ALBIFRONS*   (Humboldt 1811)

GEOG DIST: Central and Southern America BDY WT: 1150–3320g♂
HD/BDY LTH: 320–565 mm♂; 320–480 mm♂ TAIL LTH: 340–560 mm♂; 290–510 mm♂
COAT COL: light brown or cinnamon

SYNONYMS

Simia hypoleuca Humboldt 1811
Cebus versicolor Pucheran 1845
Cebus leucocephalus Gray 1865
Cebus malitiosus Elliot 1909
Cebus barbatus Geoffroy 1812
Cebus unicolor Spix 1823
Cebus gracilis Spix 1823
Cebus chrysopus Lesson 1827
Cebus flavescens Gray 1865
Cebus aequatorialis Allen 1914
Simia albifrons Humboldt 1811

VERNACULAR NAMES

Weisschulteraffe
White-fronted capuchin
Cinnamon ringtail
Caiarara branco
Sai-taua
Ouavapavi
Sajou à pieds dorés
Sajou à front blanc
Schrabrackenfanfaun
Weisstirn-Kapuziner
Cai de fronte blanca

2.3.8.4.   *CEBUS NIGRIVITTATUS*   Wagner 1847

Reproduced by courtesy of Zoological Society of London

GEOG DIST: Central and South America   BDY WT: 1150–3320g♂
HD/BDY LTH: 320–565mm♂; 320–480mm♀   TAIL LTH: 340–560 mm♂ : 290–510 mm♀
COAT COL: light or dark brown

SYNONYMS

Simia capucina Audebert 1797
Cebus olivaceus Schomburgk 1848
Cebus apiculatus Elliot 1907
Cebus griseus F. Cuvier 1819 (1824)
Cebus castaneus I. Geoffroy 1851
Cebus pucherani Dahlbom 1856
Cebus paraguayanus Reichenbach 1862
Cebus subcristatus Gray 1865
Cebus annelatus Gray 1865
Cebus capillatus Gray 1865

2.3.8.4.　*CEBUS NIGRIVITTATUS*　Wagner 1847

Brauner Kapuziner
Weeping capuchin
Sajou brun
Weeper capuchin
Yurac-machin
Matchi
Caiarara da serra
Cai capuchino
Chirimbabo
Singe pleureur
Sajou sai
Singe musqué

Cebus nigrivittatus

C. albifrons

MAP 19. Geographic distribution of different species of *Cebus* (redrawn and adapted from Hill, 1960)

MAP 20. Geographic distribution of different species of *Cebus* (redrawn and adapted from Hill, 1960)

2.3.9.1.   *ATELES PANISCUS*   (Linnaeus 1758)

Reproduced by courtesy of San Diego Zoo

GEOG DIST: South America   BDY WT: 5470–6890g♂; 5825g♀
HD/BDY LTH: 370–590mm♂; 345–660mm♀   TAIL LTH: 630–825 mm♂; 610–920 mm♀
COAT COL: black

SYNONYMS

Simia paniscus Linnaeus 1758
Sapajou paniscus Lacépède 1799
Ateles pentadactylus Geoffroy 1806
Simia chamek Humboldt 1811
Ateles ater Couvier 1823
Ateles longimembris Allen 1914
Ateles subpentadactylus Desmarest 1820
Ateles niger Noguchi 1919

2.3.9.1.   *ATELES PANISCUS*   (Linnaeus 1758)

VERNACULAR NAMES

Schwarzer Klammeraffe
Red-faced spider monkey
Schwarzgesichtsklammeraffe
Black-faced spider monkey
Black spider monkey
Chamek
Coata negro
Coaita noir

Reproduced by courtesy of Chessington Zoo

GEOG DIST: South America
HD/BDY LTH:
COAT COL: black or brownish-black

BDY WT: 5470–6890g♂; 5825g♀
TAIL LTH: > Hd/Bdy Lth

SYNONYMS

Ateles robustus Allen 1914
Ateles dariensis Goldman 1915

VERNACULAR NAMES

Brown-headed spider monkey
Columbian black spider monkey

2.3.9.3. *ATELES BELZEBUTH* (E. Geoffroy 1806)

Reproduced by courtesy of W. C. Osman Hill

GEOG DIST: South America
HD/BDY LTH:
COAT COL: black or brown; generally paler underside; pale triangular patch on forehead

BDY WT:
TAIL LTH: > Hd/Bdy Lth

2.3.9.3.   *ATELES BELZEBUTH*   E. Geoffroy 1806

SYNONYMS

Ateles marginatus Geoffroy 1809
Ateles marimonda Oken 1816
Ateles fuliginosus v. Hasselt and Kuhl 1820
Cebus brissonii Fischer 1829
Ateles hybridus Geoffroy 1829
Ateles variegatus Wagner 1840
Ateles chuva Schlegel 1876
Ameranthropoides loysi Montandon 1929
Ateles bartlettii Gray 1867
Ateles problema Schlegel in Jentink 1892

VERNACULAR NAMES

Weissbauchklammeraffe
Marimonda spider monkey
Goldstirnaffe
Variegated spider monkey
Atèle chuva
White-whiskered spider monkey
Long-haired spider monkey
Mulatto monkey
Aru
Maquiçapa
Urcu maquiçapa
Quillo maquiçapa
Coata branco
Coaita à ventre blanc
Langhaarige slingeraap

2.3.9.4. *ATELES GEOFFROYI* Hasselt and Kuhl 1820

GEOG DIST: Mexico
HD/BDY LTH:
BDY WT:
TAIL LTH: > Hd/Bdy Lth
COAT COL: gold, red, buff or dark brown; hands and feet generally black

SYNONYMS

Ateles melanochir Desmarest 1820 (Sclater 1871)
Eriodes frontatus Gray 1842
Brachyteles frontatus Gray 1843
Ateles vellerosus Gray 1866
Ateles neglectus Reinhardt 1873
Ateles tricolor Hollister 1914
Ateles grisescens Gray 1866
Ateles cucullatus Gray 1866
Ateles pan Schlegel 1876
Ateles ornatus Gray 1870
Ateles rufiventris Sclater 1872

2.3.9.4.   *ATELES GEOFFROYI*   Hasselt and Kuhl 1820

Geoffroy's Klammeraffe
Black-foreheaded spider monkey
Headed spider monkey
Guatemalan spider monkey
Costa Rica ornate spider monkey
Red spider monkey
Mono colorado
Black-headed spider monkey
Mexican spider monkey
Hooded spider monkey
Atèle aux mains noires
Red-bellied spider monkey
Mono de vientre colorado

MAP 21. Geographic distribution of the different species of *Ateles* (redrawn and adapted from Hill, 1962)

192

2.3.10.1.   *BRACHYTELES ARACHNOIDES*
(Geoffroy 1806)

Reproduced by courtesy of W. C. Osman Hill

GEOG DIST: South America   BDY WT: 9500g
HD/BDY LTH: 460–630mm♂; 470–565mm♀ TAIL LTH: 650–740 mm♂; 740–800 mm♀
COAT COL: grey to brown

BRACHYTELES ARACHNOIDES
(Geoffroy 1806)

SYNONYMS

Ateles hypoxanthus Hasselt and Kuhl 1820
Brachyteles macrotarsus Spix 1823
Eriodes hemidactylus Geoffroy 1829
Ateles arachnoides Geoffroy 1806
Simia arachnoides Humboldt 1812
Eriodes arachnoides Geoffroy 1829
Eriodes tuberifer Geoffroy 1829
Cebus hypoxanthus Fischer 1829
Cebus arachnoides Fischer 1829
Ateles hemidactilus Boitard 1845
Brachuteleus arachnoides Elliot 1913
Ateles eriodes Brehm 1876

VERNACULAR NAMES

Spinnenaffen
Brown woolly spider monkey
Brachytèle aracnoide
Atélé araignée
Eroide
Woolly spider monkey
Yellow spider monkey
Fawn-coloured Coaita

2.3.11.1. *LAGOTHRIX LAGOTRICHA* (Humboldt 1812)

Reproduced by courtesy of Oregon Regional Primate Research Center

GEOG DIST: South America   BDY WT: 3600–10,000g♂; 5000–6500g♀
HD/BDY LTH: 415–570mm♂; 390–580mm♀   TAIL LTH: 560–690 mm♂; 600–730 mm♀
COAT COL: brown or grey to blackish; face black or brownish

2.3.11.1. *LAGOTHRIX LAGOTRICHA* (Humboldt 1812)

Cebus canus Fischer 1829
Simia flavicanda Humboldt 1812
Simia cana Humboldt 1812
Lagothrix humboldtii Geoffroy 1812
Gastrimargus olivaceus Spix 1823
Lagothrix canus Geoffroy 1812
Lagothrix infumata (Spix 1823) (Wagner 1840)
Lagothrix ubericola Elliot 1909
Lagothrix poppigii Schinz 1844
Lagothrix castelnaui Geoffroy and Deville 1848
Lagothrix thomasi Elliot 1909
Lagothrix lugens Elliot 1907
Lagothrix (Orsonax) hendeei Thomas 1927
Simia lagotricha Humboldt 1812
Gastrimargus infumatus Spix 1823
Cebus lagothrix Fischer 1829
Lagothrix caparro Lesson 1840
Lagothrix infumatus Boitard 1845
Lagothrix lagothrix Forbes 1894
Lagothrix caroarensis Lönnberg 1931
Lagothrix cana Schinz 1844
Lagothrix geoffroyi Pucheran 1857

VERNACULAR NAMES

Grauer Wollaffe
Capparo
Gray barrigudo
Lagotriche de Humboldt
Brauner Wollaffe
Brown woolly monkey
Macaco-guazu
Humboldts Wollaffe

Schieferaffe
Wolaap van Humboldt
Humbolat's woolly monkey
Caparro
Barrigudo camùn
Macaco barrigudo
Barrigudo andino

2.3.11.2. *LAGOTHRIX FLAVICAUDA* (Humbolt 1812)

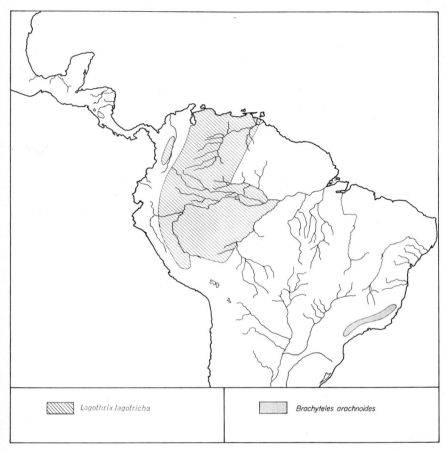

Lagothrix lagotricha

Brachyteles arachnoides

Map 22. Geographic distribution of *Brachyteles* and *Lagothrix* (redrawn and adapted from Hill, 1962)

3.1.1.1.   *MACACA SILENUS*   (Linnaeus 1758)

Reproduced by courtesy of Giardino Zoologico Torino

GEOG DIST: Southern India   BDY WT: 6755g♂
HD/BDY LTH: 510–610mm♂; 460mm♀   TAIL LTH: 255–390 mm♂; 255–320 mm♀
COAT COL: black, large grey ruff encircling face

SYNONYMS

Cercopithecus veter Erxleben 1777
Simia veter Kerr 1792
Simia ferox Shaw 1793

VERNACULAR NAMES

Wanderu
Bartaffe
Lion-tailed macaque
Ouanderu

3.1.1.2.   *MACACA NIGRA*   (Desmarest 1822)

GEOG DIST: South East Asia

HD/BDY LTH: 520–800 mm♂; 500–610 mm♀

COAT COL: black or very dark brown

BDY WT: 10,430–11,200g♂; 5100–7700g♀

TAIL LTH: 10–20 mm

SYNONYMS

Cynopithecus niger Desmarest 1821
Inuus niger Wagner 1840
Papio niger Temminck 1847
Papio nigrescens Temminck 1847
Macaca jambicus
Macaca nigrescens

VERNACULAR NAMES

Schopfipavian
Black ape
Cynopithéque niger

Celebes black ape
Celebes ape
Crested Celebes macaque

Reproduced by courtesy of Chessington Zoo

GEOG DIST: North Africa   BDY WT: 11,145 g♂
HD/BDY LTH: 559–620 mm♂; 600 mm♀   TAIL LTH: almost absent
COAT COL: black and yellow, giving mottled yellow-grey effect

3.1.1.3. *MACACA SYLVANA* (Linnaeus 1758)

SYNONYMS

Inuus ecaudatus Geoffroy 1812
Magus sylvanus Lesson 1827

VERNACULAR NAMES

Barbary ape
Magot commun
Gibraltar ape

3.1.1.4.   *MACACA ARCTOIDES*   I. Geoffroy 1831

Reproduced by courtesy of Leipzig Zoological Garden

GEOG DIST: South East Asia   BDY WT:
HD/BDY LTH: 550–700 mm♂; 500–570 mm♀   TAIL LTH: 41–100 mm♂; 10–60 mm♀
COAT COL: dark chestnut brown

SYNONYMS

Macacus brunneus Anderson 1871
Macacus arctoides Geoffroy 1831
Macacus ursinus Gervais 1854
Macacus thibetanus Milne-Edwards 1870
Macaca speciosa Cuvier 1825

VERNACULAR NAMES

Bärenmakak
Brown stump-tailed macaque
Bear macaque
Stump-tailed macaque

MACACA MAURA (Cuvier 1823)

Reproduced by courtesy of W. C. Osman Hill

GEOG DIST: South East Asia BDY WT: 8800–10,100g♂; 5105g♀
HD/BDY LTH: 445–665 mm♂; 530–570 mm♀ TAIL LTH: 45–65 mm♂; 40–55 mm♀
COAT COL: black or dark brown

Simia cuvieri Fischer 1829
Macacus ochreatus Ogilby 1840
Macacus fuscoater Schinz 1844
Macacus inornatus Gray 1866
Macacus tonkeanus Meyer 1899
Papio (Inuus) brunescens Matschie 1901
Papio (Inuus) tonsus Matschie 1901
Macacus hecki (Matschie)

VERNACULAR NAMES

Mohrenmakak
Macaque maure
Moor macaque
Booted macaque
Heck's macaque

3.1.1.6.   *MACACA SINICA*   (Linnaeus 1771)

Reproduced by courtesy of W. C. Osman Hill

GEOG DIST: Ceylon   BDY WT: 4427–8390g♂; 5405–4310g♀
HD/BDY LTH: 440–535mm♂; 430–450mm♀ TAIL LTH: 550–620 mm♂; 465–570 mm♀
COAT COL: gold or reddish brown; paler underparts

SYNONYMS

Cynomolgus audeberti Reichenbach 1862
Macaca pileatus Blyth 1863
Pithecus sinicus Linnaeus 1771
Simia sinica Linnaeus 1771

VERNACULAR NAMES

Ceylon-hutaffe
Toque monkey
Macaque couronné

205

3.1.1.7. *MACACA RADIATA* (E. Geoffroy 1812)

Drawing by Kostas, Turin

GEOG DIST: Southern India
HD/BDY LTH: 510–600 mm♂; 345–525 mm♀
COAT COL: grey-brown; paler underparts

BDY WT: 5670–8850g♂; 2930–4420g♀
TAIL LTH: 510–690 mm♂; 480–635 mm♀

SYNONYMS

Cercocebus radiatus Geoffroy 1812

VERNACULAR NAMES

Hutaffe
Bonnet monkey
Macaque commun
Bonnet macaque

3.1.1.8.　*MACACA CYCLOPIS*　(Swinhoe 1862)

Geog Dist: Formosa
Hd/Bdy Lth:
Coat Col: slate brown with darker limbs

Bdy Wt:
Tail Lth: almost absent

SYNONYMS

VERNACULAR NAMES

Formosmakak
Formosan rock macaque
Macaque de Formosa
Formosan macaque

3.1.1.9. *MACACA MULATTA* (Zimmermann 1780)

GEOG DIST: South East Asia   BDY WT: 5560–10, 900g♂; 4370–10 660g♀
HD/BDY LTH: 485–635 mm♂; 470–530 mm♀ TAIL LTH: 200–305 mm♂ 190–285 mm♀
COAT COL: brown with paler underparts

SYNONYMS

Cercopithecus mulatta Zimmermann 1780
Simia rhesus Audebert 1798
Simia erythraea Shaw 1800
Macaca cinops Hodgson 1840
Macacus lasiotis Gray 1868
Macacus vestitus Milne-Edwards 1892
Macacus rhesus True 1894
Pithecus littoralis Elliot 1909
Pithecus brevicaudatus Elliot 1913

VERNACULAR NAMES

Rhesus-Affe
Rhesus macaque
Macaque rhésus
Bandar
Rhesus monkey

208

3.1.1.10.   *MACACA FUSCATA*   (Blyth 1875)

GEOG DIST: Japan   BDY WT: 11, 100–18, 000g♂; 8300–16, 300g♀
HD/BDY LTH: 535–610mm♂; 470–600mm♀ TAIL LTH: 80–125 mm♂; 70–100 mm♀
COAT COL: yellowish brown

SYNONYMS

Inuus speciosus Temminck 1842

VERNACULAR NAMES

Japanischer Makak
Rotgesichtsmakak
Japanese macaque
Macaque à face rouge

3.1.1.11.   *MACACA NEMESTRINA*   (Linnaeus 1766)

Reproduced by courtesy of Zoological Society of London

GEOG DIST: South East Asia
HD/BDY LTH: 495–595mm♂; 465–565 mm♀
COAT COL: dark brown, paler underparts

BDY WT: 6245–14 500g♂; 4655–10 900g♀
TAIL LTH: 160–245 mm♂; 140–195 mm♀

SYNONYMS

Simia carpolegus Raffles 1822
Macacus brachyurus Smith 1842
Macacus leoninus Blyth 1863
Macacus andamanensis Bartlett 1869
Macacus pagensis Miller 1903
Macaca adusta Miller 1906
Macaca insulana Miller 1906

3.1.1.11. *MACACA NEMESTRINA*   (Linnaeus 1766)

VERNACULAR NAMES

Schweinsaffe
Löwenmakak
Pig-tailed macaque
Macaque maimon
Singe à queue de cochon
Berok
Giant rhesus
Maimon

Reproduced by courtesy of W. C. Osman Hill

3.1.1.12.   *MACACA FASCICULARIS*   Raffles 1821

GEOG DIST: South East Asia

BDY WT: 3500–8285g♂; 2500–5680g♀

HD/BDY LTH: 410–650mm♂; 385–505mm♀   TAIL LTH: 435–655mm♂; 400–545mm♀

COAT COL: pale yellow-brown or grey to dark brown; paler underparts

SYNONYMS

Simia cynomolgus Schreber 1775
Simia fascicularis Raffles 1821
Macacus carbonarius Cuvier 1825
Semnopithecus kra Lesson 1830
Macacus aureus Geoffroy 1834
Macacus umbrosus Miller 1902
Pithecus alacer Elliot 1909
Pithecus validus Elliot 1909
Pithecus mandibularis Elliot 1909
Pithecus bintangensis Elliot 1909
Macaca phylippinensis Geoffroy 1843
Macaca mindora (Hollister 1913)
Macacus palpebrosus Geoffroy 1851
Macaca mindanensis (Mearns 1905)
Macaca sulvensis (Mearns 1905)
Macaca cagayana (Mearns 1905)

VERNACULAR NAMES

Javaneraffe
Crab-eating macaque
Macaque de Buffon
Common macaque
Java monkey
Kra monkey

3.1.1.13. *MACACA ASSAMENSIS* (M'Clelland 1839)

Reproduced by courtesy of Zoologischer Garten, Berlin

GEOG DIST: South East Asia   BDY WT: 10 445–12 710g♂
HD/BDY LTH: 560–650 mm♂; 530–680 mm♀   TAIL LTH: 190–350 mm♂; 235–365 mm♀
COAT COL: yellowish brown to dark brown

SYNONYMS

Macacus problematicus Gray 1870
Macacus rhesosimilis Sclater 1872

3.1.1.13. *MACACA ASSAMENSIS* M'Clelland 1839

VERNACULAR NAMES

Assamrhesus
Assamese macaque
Macaque d'Assam
Himalayan macaque
Montane rhesus

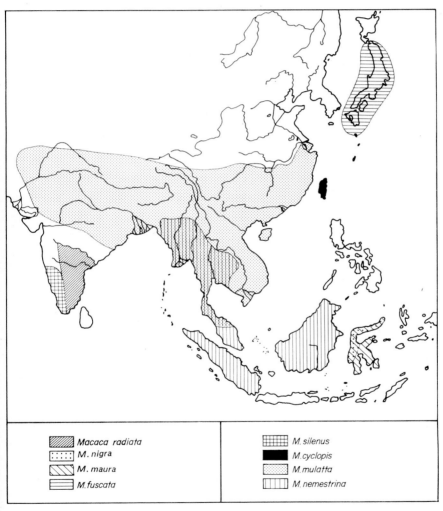

MAP 23. Geographic distribution of different species of *Macaca* (redrawn and adapted from Fiedler, 1956 and Napier, 1967)

215

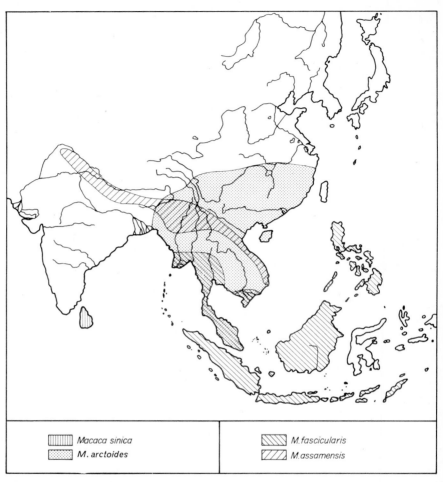

MAP 24. Geographic distribution of different species of *Macaca* (redrawn and adapted from Fiedler, 1956 and Napier, 1967)

216

3.1.2.1.   *PAPIO HAMADRYAS*   (Linnaeus 1758)

Reproduced by courtesy of Bernhard Grzimek

GEOG DIST: East Ethiopia, North Somalia,   BDY WT: 20,000–30,000g♂; 10,000–
   South West Arabia   15,000g♀
HD/BDY LTH: *ca* 800 mm   TAIL LTH: *ca* 150 mm
COAT COL: greyish; adult♂ with copious mane of light-coloured hair

SYNONYMS

Simia cynomolgos Linnaeus 1758
Hamadryas choeropothecus Lesson 1840
Papio arabicus Thomas 1900
Papio brockmanni Elliot 1909

VERNACULAR NAMES

Mantelpavian
Hamadryas baboon
Papion hamadryas
Sacred baboon

217

Reproduced by courtesy of Zoologischer Garten, Basel

3.1.2.2.   *PAPIO URSINUS*   Kerr 1792

GEOG DIST: South Africa; North to Zambia,
   Angola, Mozambique   BDY WT:
HD/BDY LTH: TAIL LTH:
COAT COL: deep brown

SYNONYMS

Simia porcaria Boddaert 1787
Cynocephalus ursinus Wagner 1840
Papio comatus Geoffroy 1812

VERNACULAR NAMES

Barenpavian
Chacma baboon
Papion chacma
Pig-tailed baboon

Reproduced by courtesy of W. C. Osman Hill

GEOG DIST: Sierra Leone, Central Ethiopia, BDY WT: 20,000–30,000g ♂; 11,000–
  North Tanzania   15,000♀
HD/BDY LTH: 735–785 mm♂; 560–666 mm♀ TAIL LTH: 520–600 mm ♂; 415–530 mm♀
COAT COL: brown

SYNONYMS

Simia anubis Fischer 1829   *Papio graueri* Lorenz 1917
Papio heuglini Matschie 1898   *Papio neumanni* Matschie 1897
Cynocephalus olivaceus Geoffroy   *Papio doguera* (Pucheran 1856)
Papio toth-ibeanus Thomas 1893

VERNACULAR NAMES

Atbarapavian   Doguera baboon
Grauer Babuin   Papion anubis
Zwergpavian   Olive baboon
Anubis baboon

220

3.1.2.4.   *PAPIO CYNOCEPHALUS*   Linnaeus (1766)

Reproduced by courtesy of Åke Nordh, Nordiska, Museet Och Skansen, Stockholm

GEOG DIST: North Angola, North
   Mozambique, East Kenya, South East
   Somalia
HD/BDY LTH: 508–1143 mm
COAT COL: brown

BDY WT: 22–50kg♂; 11–30kg♀
TAIL LTH: 456–711 mm

SYNONYMS

Cynocephalus babouin Desmarest 1820
Cynocephalus thoth Ogilby 1843
Cercopithecus ochreaceus Peters 1853
Cynocephalus langheldi Matschie 1892
Papio strepitus Elliot 1907

VERNACULAR NAMES

Gelber Babuin
Yellow baboon
Babouin cynocéphale
Long-legged baboon
Nyani
Ebbaku
Galder
Chebioyiet

Nkerebe
Abula
Nkuka
Nkobe
Abim
Echom
Lore

3.1.2.5.   *PAPIO PAPIO*   (Desmarest 1820)

GEOG DIST: Senegal to Guinea   BDY WT:
HD/BDY LTH: TAIL LTH:
COAT COL: reddish

SYNONYMS

Cynocephalus papio Desmarest 1820
Cynocephalus olivaceus Geoffroy 1851
Papio rubescens Temminck 1853

VERNACULAR NAMES

Roter Pavian
Guinea pavian
Guinea baboon
Western baboon
Babouin de Guinée

3.1.2.6. *PAPIO SPHINX* (Linnaeus 1758)

Reproduced by courtesy of W. C. Osman Hill

GEOG DIST: Central Africa
HD/BDY LTH: 810 mm♂
COAT COL: dark brown to charcoal grey; fringe of orange and yellow

BDY WT: 19 525g♂
TAIL LTH: 70 mm♂

3.1.2.6.   *PAPIO SPHINX*   (Linnaeus 1758)

Simia maimon Linnaeus 1766
Papio planirostris Elliot 1909

VERNACULAR NAMES

Mandrill
Rib-faced-baboon

3.1.2.7.   *PAPIO LEUCOPHAEUS*   (Cuvier 1807)

Reproduced by courtesy of J. Klages

GEOG DIST: Central Africa
HD/BDY LTH: 700 mm♂
COAT COL: olive green

BDY WT:
TAIL LTH: 120 mm♂

SYNONYMS

Mormon drill Lesson 1840

VERNACULAR NAMES

Drill

3.1.3.1. *THEROPITHECUS GELADA* (Rüppell 1835)

Reproduced by courtesy of San Diego Zoo

GEOG DIST: Ethiopia   BDY WT: *ca* 20 500g♂; 13 620g♀
HD/BDY LTH: 690–740mm♂; 500–650mm♀ TAIL LTH: 460–500 mm♂ ;325–410 mm♀
COAT COL: dark brown; forearms, hands, feet almost black; face dark brown, white
  eyelids

3.1.3.1. *THEROPITHECUS GELADA* (Rüppell 1835)

SYNONYMS

Macacus gelada Rüppell 1835
Theropithecus senex Pucheran 1857
Theropithecus obscurus Heuglin 1853

VERNACULAR NAMES

Blutbrustpavian
Gelada baboon
Théropithèque gélada
Bleeding heart baboon
Nacktbrustpavian
Dschelada

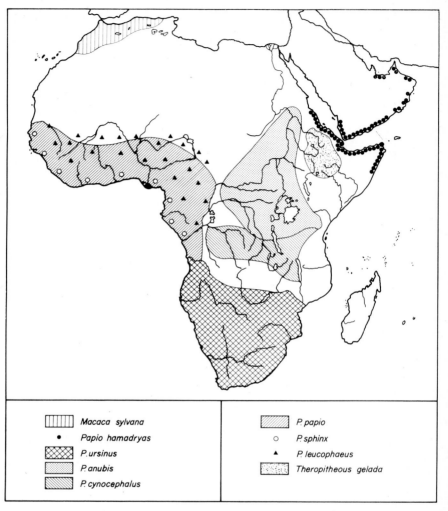

MAP 25. Geographic distribution of *Macaca sylvana*, *Theropithecus* and the different species of *Papio* (redrawn and adapted from Fiedler, 1956)

Legend:

- Macaca sylvana
- Papio hamadryas
- P. ursinus
- P. anubis
- P. cynocephalus
- P. papio
- P. sphinx
- P. leucophaeus
- Theropitheous gelada

Reproduced by courtesy of San Diego Zoo

GEOG DIST: Congo, Uganda, Tana River   BDY WT:
  (Kenya)
HD/BDY LTH: 515–580 mm♂; 440–520 mm♀   TAIL LTH: 690–785 mm♂; 590–695 mm♀
COAT COL: mushroom

3.1.4.1.　*CERCOCEBUS GALERITUS*　Peters 1879

SYNONYMS

Cercocebus agilis Milne-Edwards 1866
Cercocebus chrysogaster Lydekker 1900
Cercocebus fumosus Matschie 1914

VERNACULAR NAMES

Haubenmangabe
Crested mangabey
Mangabey à crête
Mangabey à ventre doré
Cercocèbe agile
Cercocèbe à crête
Plain-headed mangabey
Tana River mangabey

3.1.4.2.   *CERCOCEBUS ATYS*   (Audebert 1797)

Reproduced by courtesy of Tierbilder Okapia

GEOG DIST: West Africa   BDY WT:
HD/BDY LTH: 515–580 mm♂; 440–520 mm♀   TAIL LTH: 690–785 mm♂; 590–695 mm♀
COAT COL: dark smokey grey

SYNONYMS

Simia atys Audebert 1797
Cercocebus fuliginosus E. Geoffroy 1812
Cercopithecus lunulatus Temminck 1853

VERNACULAR NAMES

Mohrenmangabe   Sooty mangabey
Mangabé enfumé   White-crowned mangabey
Rauchgrave Mangabe

3.1.4.3. *CERCOCEBUS TORQUATUS* (Kerr 1792)

Reproduced by courtesy of Giardino Zoologico, Roma

GEOG DIST: West Africa   BDY WT:
HD/BDY LTH: 515–580 mm♂; 440–520 mm♀ TAIL LTH: 690–785 mm♂; 590–695 mm♀
COAT COL: dark with crown chestnut red

SYNONYMS

Simia aethiops torquatus Kerr 1792
Cercocebus collaris Gray 1843

3.1.4.3. *CERCOCEBUS TORQUATUS* (Kerr 1792)

Rotkopfmangabe
Halsbandmangabe
Weissscheitelmangabe
White-collared mangabey
Mangabey à collier blanc
Mangabey éthiops
Cercocèbe à collier blanc

3.1.4.4. *CERCOCEBUS ATERRIMUS* (Oudemans 1890)

Reproduced by courtesy of Zoologischer Garten, Basel

GEOG DIST: Congo
HD/BDY LTH: 540–615 mm♂; 435–580 mm♀
COAT COL: black

BDY WT:
TAIL LTH: 820–940 mm♂; 740–895 mm♀

3.1.4.4. *CERCOCEBUS ATERRIMUS* (Oudemans 1890)

SYNONYMS

Cercocebus hamlyni Pocock 1906

VERNACULAR NAMES

Schwarze Schopfmangabe
Black mangabey
Mangabey noir
Kuif-mangabey
Le corcocèbe noir
Mangabé huppé
Peaked mangabey

3.1.4.5. *CERCOCEBUS ALBIGENA*   (Gray 1850)

GEOG DIST: Congo and Uganda   BDY WT: 7–11kg
HD/BDY LTH: 540–615 mm♂; 435–580 mm♀ TAIL LTH: 820–940 mm♂; 740–895 mm♀
COAT COL: black

3.1.4.5.   *CERCOCEBUS ALBIGENA*   (Gray 1850)

Presbytis albigena Gray 1850
Semnocebus albigena Lydekker 1900

VERNACULAR NAMES

Grauwangen Mangabe
Mantelmangabe
Gray-cheeked mangabey
Mangabey à gorge blanche
Mangabey albigène
Black mangabey
Cercocèbe à gorge blanc
Crested mangabey
Kuif-mangabey
Mangabé huppé
Sserwagabo
Ngari
Kigari

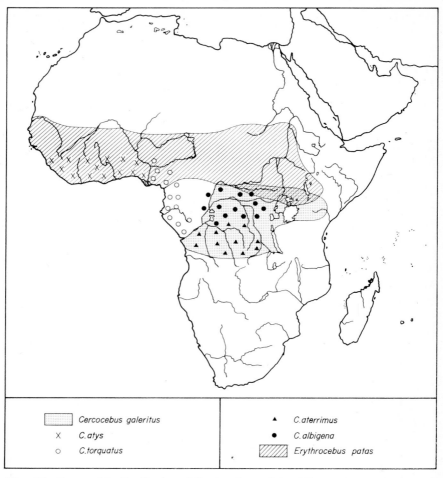

MAP 26. Geographic distribution of *Erythrocebus* and the different species of *Cercocebus* (redrawn and adapted from Fiedler, 1956 and Napier, 1967)

3.1.5.1. *CERCOPITHECUS AETHIOPS* Linnaeus 1758

Reproduced by courtesy of Tierbilder Okapia

GEOG DIST: South Africa, Senegal, Ethiopia,
　Sudan
BODY WT: 5–9kg
HD/BDY LTH: 410–645 mm♂; 315–520 mm♀ TAIL LTH: 545–1090 mm♂; 480–1020 mm
COAT COL: greenish or greyish; face sooty black

3.1.5.1. *CERCOPITHECUS AETHIOPS* Linnaeus 1758

Cercopithecus griseoviridis Desmarest 1820
Cercopithecus griseus Cuvier 1824
Chlorocebus toldti Wettstein 1916
Lasiopyga pygerythra callida Hollister 1912
Cercopithecus centralis Neumann 1900
Cercopithecus rufo-viridis Geoffroy 1842
Cercopithecus callithrichus Geoffroy 1851
Cercopithecus tantalus Ogilby 1841
Cercopithecus viridis Schultze 1910
Simia subviridis Cuvier 1821
Cercopithecus canoviridis Gray 1843
Cercopithecus cinereo-viridis Gray 1843
Cercopithecus sabaeus I. Geoffroy 1850
Cercopithecus engythithia Horsfield 1851
Cercopithecus ellenbecki Neumann 1902
Chlorocebus cailliaudi Wettstein 1918
Lasiopyga weidholzi Lorenz 1922

VERNACULAR NAMES

Grüne Meerkatze
Savannah monkey
Singe vert
Malbrouk
Grivet
Green monkey
Abellen
Abu-lang
Alesteo
Tota
Cercopithèque d'Ethiopie
Cercopiteco grigioverde
Vervet monkey
Tumbili
Serwagaba
Nkende
Ongera
Edokolet
Kamale
Suboltit
Girengwa

3.1.5.2. *CERCOPITHECUS CYNOSUROS* (Scopoli 1786)

Reproduced by courtesy of W. C. Osman Hill

3.1.5.2. *CERCOPITHECUS CYNOSUROS* (Scopoli 1786)

Geog Dist: South and Central Africa   Bdy Wt:
Hd/Bdy Lth: 410–645 mm♂; 315–520 mm♀ Tail Lth: 575–1090 mm♂; 480–1020 mm♀
Coat Col: greenish or greyish; tip of tail black, patch of reddish hairs at root and
   anus

SYNONYMS

Simia cynosuros Scopoli 1786
Simia pygerythra F. Cuvier 1821
Cercopithecus pygerythraeus Desmarest 1822
Simia erythropyga G. Cuvier 1829
Cercopithecus lalandii I. Geoffroy 1843

VERNACULAR NAMES

Vervet monkey
Pusi
Nkau
Tumbili
Ngedere
Cercopithèque vervet
Cercopiteco giallo-verde

3.1.5.3. *CERCOPITHECUS SABAEUS* (Linnaeus 1758)

Reproduced by courtesy of Jürg Klages

GEOG DIST: Central and South Africa   BDY WT:
HD/BDY LTH: 410–645 mm♂;315–520 mm♀ TAIL LTH: 575–1090 mm♂;480–1020 mm♀
COAT COL: greenish or greyish

SYNONYMS

Simia sabaea Linnaeus 1766
Cercopithecus werneri I. Geoffroy 1850
Cercopithecus callithrichus I. Geoffroy 1851

243

3.1.5.3. *CERCOPITHECUS SABAEUS* (Linnaeus 1758)

VERNACULAR NAMES

Cercopithèque Callitriche
Green monkey
Savannah monkey
Singe vert
Grünmeerkatze
Gelbgrünmeerkatze

3.1.5.4. *CERCOPITHECUS CEPHUS* (Linnaeus 1758)

Reproduced by courtesy of Jürg Klages

GEOG DIST: Gabon and Congo   BDY WT:
HD/BDY LTH: 410–645 mm♂; 315–520 mm♀ TAIL LTH: 575–1090 mm♂; 480–1020 mm♀
COAT COL: transverse pale blue shape on lips; yellow whiskers

3.1.5.4. *CERCOPITHECUS CEPHUS* (Linnaeus 1758)

SYNONYMS

Cercopithecus buccalis Leconte 1857
Cercopithecus pulcher Lorenz 1915
Simia cephus Linnaeus 1758
Cercopithecus cephus cephodes Pocock 1907
Cercopithecus inobservatus Elliot 1910
Lasiopyga cephus Elliot 1913
Lasiopyga cephodes Elliot 1913
Lasiopyga inobservata Elliot 1913
Cercopithecus cephus cephus Schwarz 1928

VERNACULAR NAMES

Blaumavemeerkatze
Moustached guenon
Red-eared guenon
Moustac de Buffon
Moustached monkey
Schnurrbartaffe
Blaumaul
Ossok

3.1.5.5. *CERCOPITHECUS DIANA*   Linnaeus 1758

Reproduced by courtesy of Tierbilder Okapia

GEOG DIST: West Africa BDY WT:
HD/BDY LTH: 410–645 mm♂;315–520 mm♀ TAIL LTH: 575–1090 mm♂;480–1020 mm♀
COAT COL: face black, white brow band, long white beard

247

3.1.5.5. *CERCOPITHECUS DIANA* Linnaeus 1758

SYNONYMS

Simia faunus Linnaeus 1766
Simia rolloway Schreber 1774
Cercopithecus dryas Schwarz 1932

VERNACULAR NAMES

Dianameerkatze
Diana monkey
Cercopithèque diane
Dianamarekatt

3.1.5.6. *CERCOPITHECUS L'HOESTI* (Sclater 1889)

Reproduced by courtesy of E. Gunther

3.1.5.6. *CERCOPITHECUS L'HOESTI* (Sclater 1889)

<small>Geog Dist</small>: Central and East Africa   <small>Bdy Wt</small>:
<small>Hd/Bdy Lth</small>: 410–700mm♂; 315–520mm♀ <small>Tail Lth</small>: 575–1090mm♂ 480–1020mm♀
<small>Coat Col</small>: limbs distal to elbows and knees black; belly blackish; whiskers white,
cheeks lighter

SYNONYMS

Cercopithecus thomas Matschie 1905
Cercopithecus insolitus Elliot 1909
Lasiopyga insolita Elliot 1913
Lasiopyga thomasi Elliot 1913
Cercopithecus thomasi Lorenz 1915

VERNACULAR NAMES

Vollbartvemeerkatze
L'Hoest's guenon
Cercopithèque de l'Hoest
Mountain guenon
L'Hoest's monkey
Mountain monkey
Enkende
Cercopithèque à barbe en collier
Nyaluasa
Engende
Embeya

3.1.5.7. *CERCOPITHECUS PREUSSI* (Matschie 1898)

GEOG DIST: Cameroun BDY WT:
HD/BDY LTH: 410–645 mm♂; 315–520 mm♀ TAIL LTH: 575–1090 mm♂; 480–1020 mm♀
COAT COL: limbs distal to elbows and knees black; belly blackish; whiskers white, cheeks greyer

SYNONYMS

Cercopithecus crossi Forbes 1905
Lasiopyga preussi Elliot 1913

VERNACULAR NAMES

Cross's monkey
Preuss's monkey

3.1.5.8. *CERCOPITHECUS HAMLYNI* Pocock 1907

Reproduced by courtesy of San Diego Zoo

GEOG DIST: Congo   BDY WT:
HD/BDY LTH: 415–645 mm♂; 315–520 mm♀   TAIL LTH: 575–1090 mm♂; 480–1020 mm♀
COAT COL: median vertical white streak on nose

SYNONYMS

Rhinostigma hamlyni Elliot 1913
Cercopithecus leucampyx Thomas et Wroughton 1910

VERNACULAR NAMES

Hamlynmeerkatze
Owl-faced guenon
Cercopithèque à tête de hibou
Hamlyn's guenon
Hamlyn's monkey
Owl-faced monkey
Cercopithèque d'Hamlyn
Eulenkopfaffe

3.1.5.9. *CERCOPITHECUS MONA*   (Schreber 1774)

Reproduced by courtesy of Zoological Society of London

GEOG DIST: West and Central Africa   BDY WT: 3–6kg
HD/BDY LTH: 415–645mm♂; 315–520mm♀ TAIL LTH: 575–1090mm♂; 480–1020mm♂
COAT COL: muzzle flesh-coloured

SYNONYMS

Simia mona Schreber 1774
Simia monacha Schreber 1804
Lasiopyga mona Elliot 1913

VERNACULAR NAMES

Monameerkatze
Mona guenon
Cercopithèque mone
Mona monkey
Mammbe
N'semgui
Mbenge

3.1.5.10. *CERCOPITHECUS CAMPBELLI* (Waterhouse 1838)

Reproduced by courtesy of W. C. Osman H

GEOG DIST: Central Africa   BDY WT:
HD/BDY LTH: 415–645 mm♂; 315–520 mm♀   TAIL LTH: 575–1090 mm♂; 480–1020 mm♀
COAT COL: muzzle flesh-coloured

SYNONYMS

Cercopithecus burnetti Gray 1842
Cercopithecus monella Gray 1870
Lasiopyga campbelli Elliot 1913

VERNACULAR NAMES

Campbell's monkey
Le mone de Campbell

3.1.5.11.   *CERCOPITHECUS WOLFI*   Meyer 1891

GEOG DIST: Central Africa
HD/BDY LTH: 415–645 mm♂; 315–520 mm♀
COAT COL: muzzle flesh-coloured

BDY WT:
TAIL LTH: 575–1090 mm♂; 480–1020 mm♀

SYNONYMS

Lasiopyga wolfi Elliot 1913

VERNACULAR NAMES

Wolf's guenon

3.1.5.12. *CERCOPITHECUS POGONIAS* (Bennett 1833)

Reproduced by courtesy of W. C. Osman Hill

GEOG DIST: Central Africa BDY WT:
HD/BDY LTH: 415–645 mm♂; 315–520 mm♀ TAIL LTH: 575–1090 mm♂; 480–1020 mm♀
COAT COL: muzzle flesh-coloured

SYNONYMS

Lasiopyga pogonias Elliot 1913

VERNACULAR NAMES

Golden-bellied monkey
Crowned guenon
Pid
Pindi
Koundi
Essouna
Poundi
Esuma
Cercopithèque pogonias

3.1.5.13. *CERCOPITHECUS DENTI* (Thomas 1907)

GEOG DIST: Central Africa   BDY WT: 3–6kg
HD/BDY LTH: 415–645 mm♂; 315–520 mm♀ TAIL LTH: 575–1090 mm♂; 480–1020 mm♀
COAT COL: muzzle flesh-coloured

SYNONYMS

VERNACULAR NAMES

Dent's guenon
La mone de Dent

3.1.5.14. *CERCOPITHECUS PETAURISTA* (Schreber 1775)

Reproduced by courtesy of Tierbilder Okapia

GEOG DIST: Central Africa

BDY WT:

HD/BDY LTH: 415–645 mm♂; 315–520 mm♀   TAIL LTH: 575–1090 mm♂; 480–1020 mm♀

COAT COL: nose spot cordate, white, yellow or red

3.1.5.14. *CERCOPITHECUS PETAURISTA* (Schreber 1775)

SYNONYMS

Simia petaurista Schreber 1775
Simia albinasus Reichenbach 1863
Cercopithecus fantiensis Matschie 1893
Cercopithecus buttikoferi Thomas 1923

VERNACULAR NAMES

Lesser white-nosed monkey

3.1.5.15.   *CERCOPITHECUS NEGLECTUS*   (Schlegel 1876)

Reproduced by courtesy of Antwerp Zoo

GEOG DIST: East and Central Africa   BDY WT: 5.8–7.8kg♂; 4.5–4.98kg♀
HD/BDY LTH: 415–645 mm♂; 315–520mm♀ TAIL LTH: 575–1090mm♂; 480–1020mm♀
COAT COL: face black with chestnut brown band, short white beard

SYNONYMS

Cercopithecus brazzae Milne-Edwards 1886
Cercopithecus leucocampyx Gray 1870
Cercopithecus brazziformis Pocock 1907
Cercopithecus ezrae Pocock 1908
Lasiopyga neglecta Elliot 1913
Cercopithecus uellensis Lönnberg 1919
Lasiopyga brazzae Allen 1925

VERNACULAR NAMES

Neglectus monkey	Cercopithèque de Brazza
Kalasinga	Schlegel's guenon
Adu	De Brazza's monkey
Ebubusi	Avut
Enyuru	Fum
Brazzameerkatze	Avout
De Brazza's guenon	Foum

3.1.5.15. *CERCOPITHECUS NEGLECTUS* (Schlegel 1876)

Foung
Pounnga
Pfoumwa
Pouhon
Pfong

3.1.5.16. *CERCOPITHECUS NICTITANS* (Linnaeus 1766)

Reproduced by courtesy of Chester Zoo

GEOG DIST: Central Africa
HD/BDY LTH: 415–645 mm♂; 315–520 mm♀
COAT COL: nasal spot yellow, oval

BDY WT:
TAIL LTH: 575–1090 mm♂; 410–1020 mm♀

SYNONYMS

Simia nictitans Linnaeus 1766
Cercopithecus sticticeps Elliot 1909
Cercopithecus büttikoferi Jentink 1886
Cercopithecus martini Waterhouse 1838
Simia albinasus Reichenbach 1863
Cercopithecus schmidti Matschie 1892
Cercopithecus signatus Jentink 1886
Lasiopyga nictitans Illiger 1811

VERNACULAR NAMES

Weissnasenmeerkatze
Spot-nosed guenon
Blanc-nez
Putty-nosed guenon
Putty-nosed monkey
Hocheur monkey
Greater white-nosed guenon
La guenon à nez blanc proéminent

3.1.5.17. *CERCOPITHECUS ASCANIUS* (Audebert 1799)

GEOG DIST: Central Africa   BDY WT: 3178–6356g♂; 1816–3859g♀
HD/BDY LTH: 415–645mm♂; 315–520mm♀ TAIL LTH: 575–1070mm♂; 480–1020mm♀
COAT COL: nasal spot yellow, oval

SYNONYMS

Simia ascanius Audebert 1799
Cercopithecus melanogenys Gray 1845
Cercopithecus histrio Reichenbach 1863
Cercopithecus picturatus Santos 1886

VERNACULAR NAMES

Cercopitheque pain-a-cacheter   Nkunga
Black-cheeked white-nosed monkey   Nkembo
Redtail monkey   Nkende
White-nosed monkey   Ikondo
Nakabugo

263

3.1.5.18.
CERCOPITHECUS ERYTHROTIS (Waterhouse 1838)

Reproduced by courtesy of Kenneth Green

GEOG DIST: Central Africa   BDY WT:
HD/BDY LTH: 415–643 mm♂; 315–520 mm♀ TAIL LTH: 545–1090 mm♂; 480–1020 mm♀
COAT COL: white nasal spot; red tail; red hairs on perineal region

SYNONYMS

Lasiopyga erythrotis Elliot 1913
Cercopithecus sclateri Pocock 1904

VERNACULAR NAMES

Red-eared nose-spotted guenon
Rotohr Meerkatze

264

3.1.5.19. *CERCOPITHECUS ERYTHROGASTER* Gray 1866

Reproduced by courtesy of W. C. Osman Hill

3.1.5.19. *CERCOPITHECUS ERYTHROGASTER* Gray 1866

GEOG DIST: Central Africa   BDY WT:
HD/BDY LTH: 415–645 mm♂; 315–520 mm♀ TAIL LTH: 575–1090 mm♂; 480–1020 mm♀
COAT COL: white nasal spot; ventral colouration faint red

SYNONYMS

Cercopithecus signatus Jentink 1886
Lasiopyga erythrogaster Elliot 1913

VERNACULAR NAMES

Red-bellied guenon
Jentink's guenon

3.1.5.20. *CERCOPITHECUS MITIS* (Wolf 1822)

GEOG DIST: Central and South Africa   BDY WT: 6–12kg
HD/BDY LTH: 415–645 mm♂; 315–520 mm♀ TAIL LTH: 575–1090 mm♂; 480–1020 mm♀
COAT COL: limbs distal to elbows and knees black; belly blackish; whiskers dark

3.1.5.20. *CERCOPITHECUS MITIS* (Wolf 1822)

SYNONYMS

Simia leucampyx Fischer 1829
Cercopithecus diadematus Geoffroy 1834
Cercopithecus pluto Gray 1848
Semnopithecus albogularis Skyes 1831
Cercopithecus albotorquates Poursargues 1896
Cercopithecus erythrarchus Peters 1852
Cercopithecus stairsi Sclater 1892
Cercopithecus kandti Matschie 1905
Cercopithecus kolbi Neumann 1902
Cercopithecus moloneyi Sclater 1893
Cercopithecus stuhlmanni Matschie 1893
Lasiopyga leucampyx Elliot 1913
Cercopithecus leucampyx Sclater 1893
Lasiopyga albigularis Elliot 1913
Cercopithecus monoides I. Geoffroy 1841
Cercopithecus rufilatus Pocock 1907

VERNACULAR NAMES

Diademmcerkatze
Diademed guenon
Cercopithèque à diadème
Cercopithèque à gorge blanche
Singe argenté
Diademed monkey
Diademaffe
Grossbärtige Meerkatze
Sanfte Meerkatze
White-throated monkey
White-throated guenon
Kima (or nchima)
Cercopithèque à collier blanc
Cercopiteco a gola bianca
Weisskehl Meerkatze

Mitis monkey
Blue monkey
Sykes monkey
Sengwa
Cheptjegayandet
Nko

3.1.5.21. *CERCOPITHECUS NIGROVIRIDIS* Pocock 1907

Reproduced by courtesy of Ph. R. van Nostrand

GEOG DIST: Congo

BDY WT:

HD/BDY LTH: 460–510 mm♂; 410 mm♀   TAIL LTH: 500–525♂; 535 mm♀

COAT COL: chin, throat, chest, belly, inner side of legs yellowish white

SYNONYMS

Lasiopyga (Chlorocebus) nigriviridis Elliot 1913
Allenopithecus nigroviridis Pocock 1907

VERNACULAR NAMES

Cercopithèque noir et vert
Allen's monkey
Singe de Pocock
Swamp monkey
Swamp guenon
Blackish-green guenon
Allen's baboon-like monkey
Sumpfmeerkatze

3.1.5.22.   *CERCOPITHECUS TALAPOIN*   (Schreber 1774)

Reproduced by courtesy of Tierbilder Okapia

GEOG DIST: West-Central Africa   BDY WT: 1230–1280g♂; 745–820g♀
HD/BDY LTH: 350 mm♂; 340–370 mm♀   TAIL LTH: 375 mm♂; 360–380 mm♀
COAT COL: back green, speckled black; yellowish limbs; orange-ringed eyes

3.1.5.22. *CERCOPITHECUS TALAPOIN* (Schreber 1774)

Cercopithecus pileatus Geoffroy 1812
Miopithecus capillatus Geoffroy 1842
Simia talapoin Schreber 1774
Simia (Cercopithecus) niger Kerr 1792
Simia melarhina F. Cuvier 1829
Cercopithecus melarhinus Schinz 1844
Miopithecus talapoin Geoffroy 1849
Cercopithecus coronatus Rode 1937

VERNACULAR NAMES

Zwergmeerkatze
Talapoin
Singe des paleturies
Ozem
Mélarhine

3.1.6.1. *ERYTHROCEBUS PATAS* (Schreber 1774)

GEOG DIST: Sub-Saharan Africa
HD/BDY LTH: 575–750 mm♂
COAT COL: red-brown

BDY WT: 7483–12 600g♂; 4082–7100g♀
TAIL LTH: 620–740 mm♂

SYNONYMS

Simia ruber Kerr 1792
Cercopithecus pyrrhonotus Hemprich and Ehrenberg 1832
Erythrocebus baumstarcki Matschie 1905
Simia rubra Gmelin 1788
Simia nigro-fasciatus Kerr 1792
Simia albo-fasciatus Kerr 1792

VERNACULAR NAMES

Husarenaffe
Red monkey
Cercopithèque patas
Military monkey
Patas monkey

Ayom
Elwala
Engabwar
Akahinda
Naggawo

272

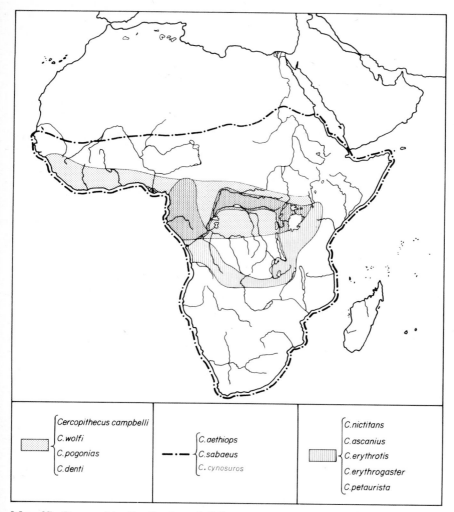

MAP 27. Geographic distribution of different species of *Cercopithecus* (redrawn and adapted from Napier, 1967)

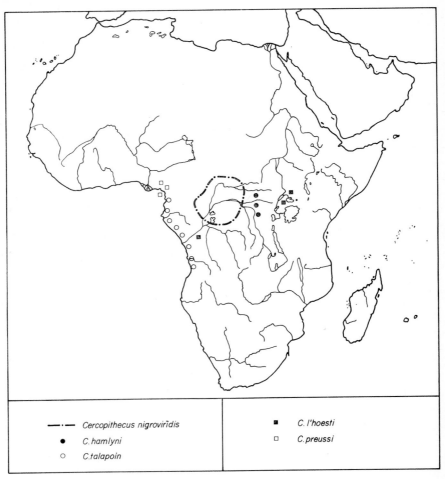

—·— *Cercopithecus nigroviridis*	■ *C. l'hoesti*
● *C. hamlyni*	□ *C. preussi*
○ *C. talapoin*	

MAP 28. Geographic distribution of different species of *Cercopithecus* (redrawn and adapted from Napier, 1967)

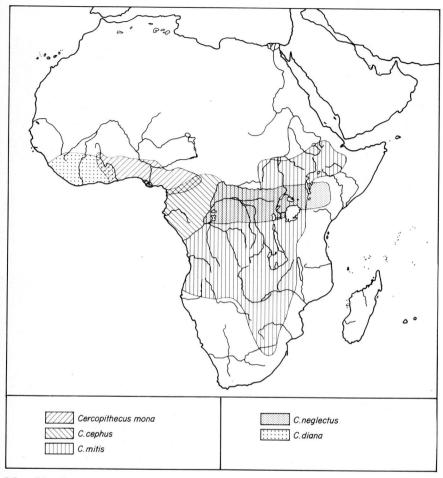

	Cercopithecus mona		C. neglectus
	C. cephus		C. diana
	C. mitis		

MAP 29. Geographic distribution of different species of *Cercopithecus* (redrawn and adapted from Napier, 1967)

3.2.1.1.   *PRESBYTIS ENTELLUS*   (Dufresne 1797)

Geog Dist: India, Pakistan, Nepal, Kashmir Bdy Wt: 9534–20 884g♂; 3178–17 706g♀
Hd/Bdy Lth: 415–787 mm♂; 432–695 mm♀ Tail Lth: 495–1092 mm♂; 599–1016 mm♀
Coat Col: adult brown, slate grey or buff

SYNONYMS

Semnopithecus schistaceus Hodgson 1840
Semnopithecus hypoleucus Blyth 1841
Semnopithecus priam Blyth 1844

276

3.2.1.1. *PRESBYTIS ENTELLUS* (Dufresne 1797)

Hulman
Hanuman
Langur
Entellus monkey
Houlemann
Entelle
Entellus langur
Hanuman langur
Hoelman
Langoor
Sacred langur
True langur

Reproduced by courtesy of Zoologischer Garten, Berlin

GEOG DIST: Extreme South West India and
    Ceylon   BDY WT: 3859–13 166g♂; 4313–11 350g♀
HD/BDY LTH: 415–587 mm♂;432–695 mm♀   TAIL LTH: 495–1092 mm♂; 599–1016 mm♀
COAT COL: black or grey; whitish whiskers

SYNONYMS

Kasi senex Erxleben 1777
Cercopithecus vetulus Erxleben 1777
Cercopithecus kephalopterus Zimmermann 1780
Cercopithecus leucoprymnus Otto 1825
Presbytis ursinus Blyth 1851

VERNACULAR NAMES

Weissbartschlankaffe
Purple-faced langur
Wanderou
Semnopithèque blanchâtre

PRESBYTIS JOHNII (Fischer 1829)

Reproduced by courtesy of San Diego Zoo

GEOG DIST: Extreme South West India,
  Ceylon
HD/BDY LTH: 415–587 mm♂; 432–695 mm♀
COAT COL: black or grey

BDY WT: 3859–13 166g♂; 4313–11 350g♀
TAIL LTH: 495–1092 mm♂; 599–1016 mm♀

SYNONYMS

Semnopithecus cucullatus
Kasi johnii Fischer 1829

VERNACULAR NAMES

Nilgiri-langur
John's langur

Semnopithèque des Nilgiris
Black langur

3.2.1.4. *PRESBYTIS AYGULA* (Linnaeus 1758)

GEOG DIST: Thailand, Malaya, Sumatra,
  Borneo, Java BDY WT 5500–7037g♂; 5500–7000g♀
HD/BDY LTH: 415–787 mm♂; 432–695 mm♀ TAIL LTH: 495–1092 mm♂; 599–1016 mm♀
COAT COL: black, grey or brown

SYNONYMS

Semnopithecus hosei Thomas 1889
Semnopithecus thomasi Collett 1892
Semnopithecus sabanus Thomas 1893
Presbytis canicrus Miller 1934

VERNACULAR NAMES

Sunda island leaf-monkey
Mitred leaf-monkey

3.2.1.5. *PRESBYTIS MELALOPHOS* (Raffles 1821)

Reproduced by courtesy of Rühmekorf, Hannover

3.2.1.5.   *PRESBYTIS MELALOPHOS*   Raffles 1821

GEOG DIST: Thailand, Malaya, Sumatra,
    Borneo, Java   BDY WT: 5500–7037g$\male$; 5500–7000g$\female$
HD/BDY LTH: 415–787 mm$\male$; 432–695 mm$\female$ TAIL LTH: 495–1092 mm$\male$; 599–1016 mm$\female$
COAT COL: black, grey or brown

SYNONYMS

Semnopithecus femoralis Martin 1838
Semnopithecus siamensis Müller and Schlegel 1841
Semnopithecus chrysomelas Müller 1838
Semnopithecus cruciger Thomas 1892
Semnopithecus natunae Thomas and Hartert 1894

VERNACULAR NAMES

Roter Schlankaffe
Banded leaf-monkey
Semnopithèque melalophe
Simpai
Cimepaye

3.2.1.6.   *PRESBYTIS RUBICUNDUS*   (Müller 1838)

GEOG DIST: Thailand, Malaya, Sumatra,
   Borneo, Java
HD/BDY LTH: 415–787 mm♂; 432–695 mm♀
COAT COL: red-brown with blue facial skin

BDY WT: 5500–7037g♂; 5500–7000g♂
TAIL LTH: 495–1092 mm♂; 599–1016 mm♀

SYNONYMS

Presbytis carimatae Miller 1906

VERNACULAR NAMES

Maroon leaf-monkey

3.2.1.7. *PRESBYTIS FRONTATUS* (Müller 1838)

GEOG DIST: Thailand, Malaya, Sumatra,
 Borneo, Java   BDY WT: 5500–7037g♂; 5500–7000g♀
HD/BDY LTH: 415–787 mm♂; 432–695 mm♀ TAIL LTH: 495–1092 mm♂; 599–1016 mm♀
COAT COL: black, grey or brown; naked patch on forehead

SYNONYMS

VERNACULAR NAMES

Weisstirnschlankaffe
White-fronted leaf-monkey
Bald leaf-monkey

3.2.1.8. *PRESBYTIS CRISTATUS* (Raffles 1821)

Reproduced by courtesy of Ph. R. van Nostrand

GEOG DIST: South East Asia   BDY WT: 3650–13 620g♂; 4767–11 350g♀
HD/BDY LTH: 415–787 mm♂; 432–695 mm♀ TAIL LTH: 495–1092 mm♂; 599–1016 mm♀
COAT COL: black

SYNONYMS

Cercopithecus auratus Geoffroy 1812
Semnopithecus pyrrhus Horsfield 1823
Trachypithecus cristatus Raffles 1821

VERNACULAR NAMES

Silvered leaf-monkey
Budeng
Crested langur
Negro langur

Reproduced by courtesy of W. C. Osman Hill

GEOG DIST: South East Asia   BDY WT: 3650–13 620g♂; 4767–11 350g♀
HD/BDY LTH: 415–787 mm♂; 432–695 mm♀ TAIL LTH: 495–1092 mm♂; 599–1016 mm♀
COAT COL: black, brown or grey; white circles round eyes; white patch on lips

SYNONYMS

VERNACULAR NAMES

Phayre's Schlankaffe
Phayre's leaf-monkey
Semnopithèque de Phayre

Reproduced by courtesy of Tierbilder Okapia

GEOG DIST: South East Asia   BDY WT: 3650–13 620g♂; 4767–11 350g♀
HD/BDY LTH: 415–787mm♂; 432–695mm♀ TAIL LTH: 495–1092mm♂; 599–1016mm♀
COAT COL: black, brown or grey; white-circled eyes; white patch on lips, hind limbs, tail and crown lighter than back

SYNONYMS

Pygathrix flavicauda Elliot 1910
Pigathrix sanctorum Elliot 1910

VERNACULAR NAMES

Rauchgrauer Blätteraffe
Dusky leaf-monkey
Semnopithèque obscur
Brillen langur
Spectacled leaf-monkey

3.2.1.11. *PRESBYTIS POTENZIANI* (Bonaparte 1856)

GEOG DIST: South East Asia   BDY WT: 3650–13 620g♂; 4767–11 350g♀
HD/BDY LTH: 415–787 mm♂; 432–695 mm♀ TAIL LTH: 495–1092 mm♂; 599–1016 mm♀
COAT COL: thick; some white on head, neck and chest

SYNONYMS

VERNACULAR NAMES

Mentawischlankaffe
Mentawi leaf-monkey

Reproduced by courtesy of Zoologischer Garten, Hannover

3.2.1.12. *PRESBYTIS FRANCOISI* (Pousargues 1898)

GEOG DIST: South East Asia   BDY WT: 3650–13 620g♂; 4767–11 350g♀
HD/BDY LTH: 415–787 mm♂; 432–695 mm♀ TAIL LTH: 495–1092 mm♂; 599–1016mm♀
COAT COL: black with variable amount of white on head

SYNONYMS

Semnopithecus poliocephalus Trouessaert 1911

VERNACULAR NAMES

François's Schlankaffe
François's leaf-monkey
Semnopithèque de François
Tonkinese lutong

3.2.1.13.   *PRESBYTIS PILEATUS*   (Blyth 1843)

Reproduced by courtesy of San Diego Zoo

GEOG DIST: South East Asia   BDY WT: 3650–13 620g♂; 4767–11 350g♀
HD/BDY LTH: 415–787 mm♂; 432–695 mm♀ TAIL LTH: 495–1092 mm♂; 599–1016 mm♀
COAT COL: grey or blackish-grey

SYNONYMS

VERNACULAR NAMES

Schopfhulman
Capped monkey
Bonneted langur

3.2.1.14.   *PRESBYTIS GEEI*   (Gee 1956)

Reproduced by courtesy of W. C. Osman Hill

GEOG DIST: South East Asia   BDY WT: 3650–13 620g♂; 4767–11 350g♀
HD/BDY LTH: 415–787 mm♂; 432–695 mm♀   TAIL LTH: 495–1092 mm♂; 599–1016 mm♀
COAT COL: cream or gold

SYNONYMS

VERNACULAR NAMES

3.2.2.1. *PYGATHRIX NEMAEUS* (Linnaeus 1771)

Reproduced by courtesy of Zoological Society of London

GEOG DIST: Laos, Vietnam, Hainam Island BDY WT:
HD/BDY LTH: 550–820 mm♂; 597–630 mm♀ TAIL LTH: 600–769 mm♂; 597–665 mm♀
COAT COL: grey with black ticking; white cheeks and neck

3.2.2.1.   *PYGATHRIX NEMAEUS*   (Linnaeus 1771)

Semnopithecus nigripes Milne-Edwards 1871

VERNACULAR NAMES

Kleideraffe
Douc langur
Painted monkey
Variegated langur
Rhinopithèque douc

3.2.3.1.
RHINOPITHECUS ROXELLANAE (Milne-Edwards 1870)

GEOG DIST: West China   BDY WT:
HD/BDY LTH: 560–830 mm♂;500–740 mm♀   TAIL LTH: 610–920 mm♂; 510–1040 mm♀
COAT COL: chocolate to dark grey; forehead, cheeks and throat buff to gold or white

3.2.3.1.
RHINOPITHECUS ROXELLANAE   (Milne-Edwards 1870)

SYNONYMS

Rhinopithecus bieti Milne-Edwards 1897
Rhinopithecus brelichi Thomas 1903

VERNACULAR NAMES

Tibet-langur
Stumpfnasenaffe
Snub-nosed monkey
Golden monkey
Rhinopithèque de Roxellane
Roxellane's monkey

3.2.3.2. *RHINOPITHECUS AVUNCULUS* Dollman 1912

GEOG DIST: North Vietnam   BDY WT:
HD/BDY LTH: 560–830 mm♂; 500–740 mm♀   TAIL LTH: 610–920 mm♂; 510–1040 mm♀
COAT COL: black; forehead, cheeks, whiskers and throat white

SYNONYMS

Presbytiscus avunculus Pocock 1926

VERNACULAR NAMES

Tonkin-Stumpfnasenaffe
Tonkin-snub-nosed monkey
Rhinopithèque de Tonkin

3.2.4.1.   *SIMIAS CONCOLOR*   Miller 1903

GEOG DIST: Mentawai Islands   BDY WT: 7150g♀
HD/BDY LTH: 490–550mm♂; 460–550mm♀ TAIL LTH: 130–190 mm♂; 100–151 mm♀
COAT COL: dark brown; hands and feet almost black

SYNONYMS

VERNACULAR NAMES

Pig-tailed monkey
Mentawi Islands langur
Pagi Islands langur
Pageh-Stumpfnasenaffe
Pig-tailed langur

Reproduced by courtesy of Tierbilder Okapia

3.2.5.1. *NASALIS LARVATUS* (Wurmb 1781)

GEOG DIST: Borneo   BDY WT: 11 700–23 608g♂; 8165–11 794g♀
HD/BDY LTH: 555–723 mm♂; 540–605 mm♀   TAIL LTH: 660–745 mm♂; 570–620 mm♀
COAT COL: red-brown, darkest on crown; dark stripe down back of neck to shoulders; cream cheeks and ruff

SYNONYMS

Simia capistratus Kerr 1792

VERNACULAR NAMES

Nasenaffe
Kahau
Proboscis monkey
Nasique
Neusaap

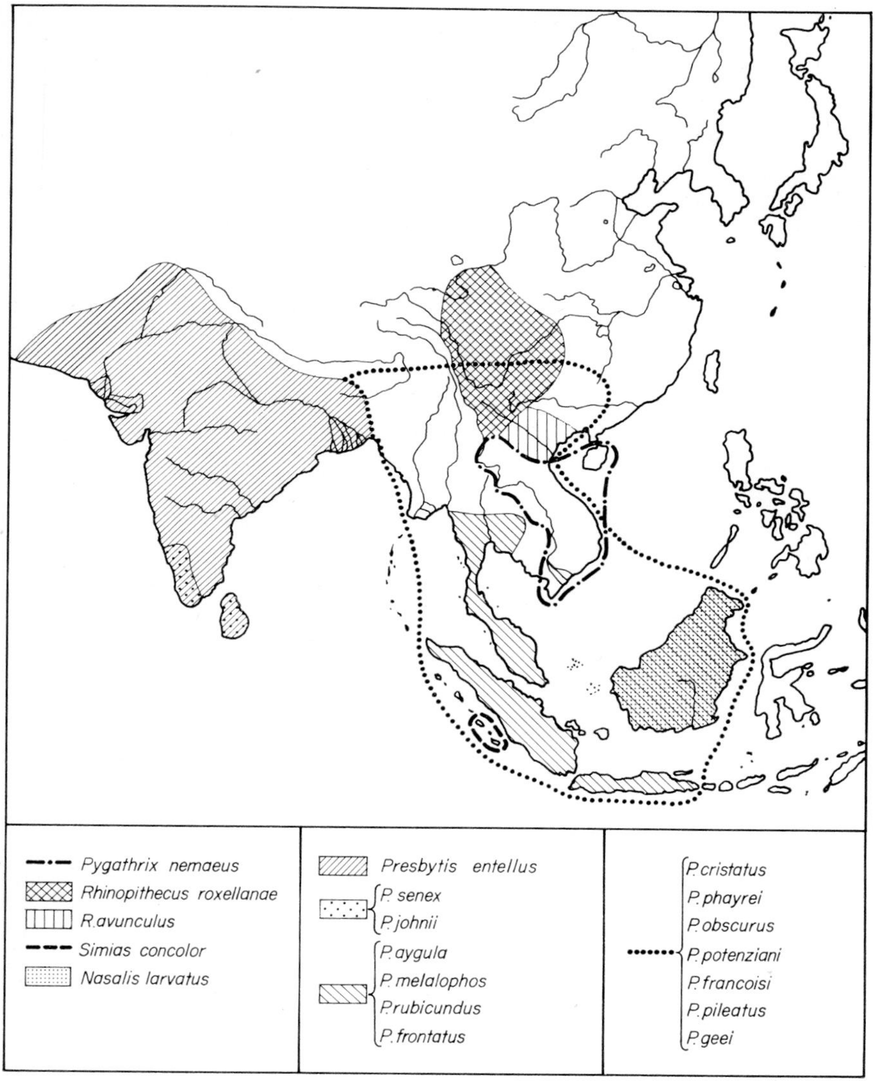
Pygathrix nemaeus
Rhinopithecus roxellanae
R.avunculus
Simias concolor
Nasalis larvatus
Presbytis entellus
P. senex
P. johnii
P. aygula
P. melalophos
P. rubicundus
P. frontatus
P. cristatus
P. phayrei
P. obscurus
P. potenziani
P. francoisi
P. pileatus
P. geei

Map 30. Geographic distribution of *Pygathrix*, *Simias*, *Nasalis* and the different species of *Rhinopithecus* and *Presbytis* (redrawn and adapted from Fiedler, 1956 and Napier, 1967)

3.2.6.1. *COLOBUS POLYKOMOS* (Zimmermann 1780)

Reproduced by courtesy of Zoological Society of London

GEOG DIST: Central Africa   BDY WT: 13–23kg
HD/BDY LTH: 490–640 mm♂; 505–610 mm♀ TAIL LTH: 720–890 mm♂; 645–880 mm♀
COAT COL: white shoulder-mantle, long white face whiskers, whorl on crown

SYNONYMS

Simia regalis Kerr 1792
Semnopithecus vellerosus Geoffroy 1834
Colobus ursinus Ogilby 1835
Colobus satanas Waterhouse 1838
Stachycolobus municus Matschie 1917

3.2.6.1. *COLOBUS POLYKOMOS* Zimmermann 1780

Weissbartstummelaffe
Bärenstummelaffe
Black and white colobus
Colobe à camail
Colobo ourson
Colobe satan
Satansaffe
Black colobus
Colobe noire
Full-bottom monkey
King colobus
Mbega
Kuluzu
Munyunga
Nkomo
Pied colobus

3.2.6.2. *COLOBUS ABYSSINICUS* (Oken 1816)

Reproduced by courtesy of Zoological Society of London

GEOG DIST: Central Africa   BDY WT: 13–23kg
HD/BDY LTH: 535–690mm♂; 485–640mm♀ TAIL LTH: 670–885 mm♂; 715–825 mm♀
COAT COL: white cape over rump, woolly white beard, black bonnet, white brush

SYNONYMS

Lemur abyssinicus Oken 1816
Guereza rüppelli Gray 1870
Colobus guereza Rüppel 1835
Guereza occidentalis Rochebrune 1886–87
Colobus angolensis Sclater 1860
Colobus palliatus Peters 1868
Colobus ruwenzorii Thomas 1901

COLOBUS ABYSSINICUS   Rüppell 1835

VERNACULAR NAMES

Abessinischer Guereza
Colobe guéréza
Mbega
Abyssinian guereza
Black and white colobus
Guereza
Mantelaffe
Mountain guereza
Weisschwanzguereza
Westlicher guereza
Western colobus
Angolaguereza
Colobe à épaules blanches
Kuluzu
Ngeye
Dolo
Ekiremu
Mongasiet
Etepes
Echu
Pied colobus

3.2.6.3.   *COLOBUS VERUS*   (Van Beneden 1838)

GEOG DIST: West Africa   BDY WT: 3300–4400g♂; 2900–4100g♀
HD/BDY LTH: 430–480 mm♂; 435–490 mm♀   TAIL LTH: 570–640 mm♂; 570–640 mm♀
COAT COL: olive-grey

SYNONYMS

Semnopithecus olivaceus Wagner 1840
Colobus cristatus Gray 1866

VERNACULAR NAMES

Schopfstummelaffe
Van Beneden's colobus
Colobe à huppe
Olive colobus
Van Beneden's monkey

COLOBUS BADIUS (Kerr 1792)

Reproduced by courtesy of Zoological Society of New York

GEOG DIST: West Africa, Congo, Uganda   BDY WT: 9–12·5kg♂; 7–9kg♀
HD/BDY LTH: 580–670mm♂; 485–620mm♀ TAIL LTH: 550–800mm♂; 412–790 mm♀
COAT COL: glossy black; chestnut red forearms, legs, underparts

3.2.6.4. *COLOBUS BADIUS* (Kerr 1792)

Simia ferruginea Shaw 1800
Colobus fuliginosus Ogilby 1835
Colobus rufomitratus Peters 1879
Colobus foai Pousargues 1899
Pitliocolobus preussi Matschie 1900

VERNACULAR NAMES

Roter Stummelaffe
Red colobus
Colobe ferrugineux, fuligineux
Brown colobus
Colobe bai
Ekajansi

3.2.6.5.   *COLOBUS KIRKII*   (Gray 1868)

Reproduced by courtesy of Photo V. Six, Antwerp Zoo

3.2.6.5. *COLOBUS KIRKII* (Gray 1868)

GEOG DIST: West and Central Africa   BDY WT:
HD/BDY LTH: 455 610 mm♂; 470–600 mm♀ TAIL LTH: 550–800 mm♂; 412–790 mm♀
COAT COL: glossy black; chestnut red forearms, legs, underparts

SYNONYMS

Procolobus kirkii Gray 1868

VERNACULAR NAMES

Kirks Stummelaffe
Kirk's colobus
Colobe de Zansibar

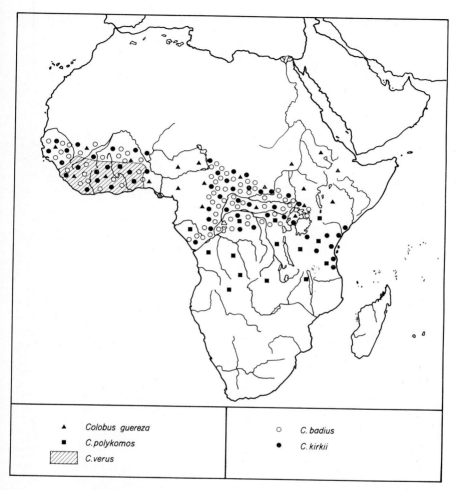

MAP 31. Geographic distribution of the different species of *Colobus* (redrawn and adapted from Fiedler, 1956 and Napier, 1967)

Legend:
▲ Colobus guereza
■ C. polykomos
▨ C. verus
○ C. badius
● C. kirkii

3.3.1.1.   *HYLOBATES LAR*   (Linnaeus 1771)

Reproduced by courtesy of Jürg Klages

GEOG DIST: Indochina, Thailand, Sumatra   BDY WT: 4300–7928g♂; 4110–6800g♀
HD/BDY LTH: 403–635mm♂; 408–622mm♀   TAIL LTH:
COAT COL: variable according to age

3.3.1.1.   *HYLOBATES LAR*   (Linnaeus 1771)

SYNONYMS

Simia longimana Schreber 1774
Pithecus variegatus Geoffroy 1812
Simia albimana Vigors and Horsfield 1828
Hylobates entelloides Geoffroy 1842
Hylobates pileatus Gray 1861

VERNACULAR NAMES

Weisshändiger Gibbon
White-handed gibbon
Gibbon à mains blanches

HYLOBATES LAR PILEATUS

Reproduced by courtesy of Giardino Zoologico Roma

GEOG DIST: Sumatra, Thailand   BDY WT: 4300–7928g♂; 4110–6800g♀
HD/BDY LTH: 403–635 mm♂; 408–622 mm♀   TAIL LTH:
COAT COL: variable according to age

SYNONYMS

Hylobates unko Lesson 1829

VERNACULAR NAMES

Unka
Dark-handed gibbon
Gibbon agile

3.3.1.3.   *HYLOBATES MOLOCH*   (Audebert 1797)

Reproduced by courtesy of C. P. Groves

GEOG DIST: Borneo   BDY WT: 4300–7928g♂; 4110–6800g♀
HD/BDY LTH: 403–635 mm♂; 408–622 mm♀ TAIL LTH:
COAT COL: grey-brown

3.3.1.3.   *HYLOBATES MOLOCH*   (Audebert 1797)

Simia moloch Audebert 1797
Pithecus leuciscus Geoffroy 1812
Hylobates mülleri Martin 1841
Hylobates funereus Geoffroy 1850

VERNACULAR NAMES

Wau wau
Sunda island gibbon
Gibbon cendré
Grey gibbon
Silver gibbon
Silvery gibbon

3.3.1.4. *HYLOBATES CONCOLOR* (Harlan 1826)

Reproduced by courtesy of Dick Hofmeister, Smithsonian Institution, Washington DC

3.3.1.4. *HYLOBATES CONCOLOR*   (Harlan 1826)

GEOG DIST: Indochina   BDY WT: 4300–7928g♂; 4110–6800g♀
HD/BDY LTH: 403–635 mm♂; 408–622 mm♀   TAIL LTH:
COAT COL: variable according to age and sex

SYNONYMS

Hylobates niger Ogilby 1840
Hylobates leucogenys Ogilby 1840
Hylobates nasutus Kunkel d'Herculais 1884
Hylobates hainanus Thomas 1892
Hylobates gabriellae Thomas 1909
Nomaseus concolor Harlan 1826

VERNACULAR NAMES

Black gibbon
White-cheeked gibbon

3.3.1.5.   *HYLOBATES HOOLOCK*   (Harlan 1834)

3.3.1.5.   *HYLOBATES HOOLOCK*   (Harlan 1834)

GEOG DIST: Assam, Burma, West Yunnam   BDY WT: 4300–7928g♂; 4110–6800g♀
HD/BDY LTH: 403–635 mm♂; 408–622 mm♀   TAIL LTH:
COAT COL: black

SYNONYMS

Hylobates fuscus Winslow Lewis 1834

VERNACULAR NAMES

Hulok
Hoolock gibbon
Gibbon houlock

3.3.1.6.   *HYLOBATES KLOSSII*   (Miller 1903)

Reproduced by courtesy of Elsbeth Siegrist

3.3.1.6.   *HYLOBATES KLOSSII*   (Miller 1903)

GEOG DIST: Mentawai Islands   BDY WT: 4300–7928g♂; 4110–6800g♀
HD/BDY LTH: 403–635mm♂; 408–622 mm♀   TAIL LTH:
COAT COL: black

SYNONYMS

Hylobates sericus Mattew and Granger 1923
Symphalangus brachytanites

VERNACULAR NAMES

Zwerg-"Siamang"
Kloss's gibbon
Dwarf gibbon
Dwarf siamang
Pygmy gibbon

3.3.2.1. *SYMPHALANGUS SYNDACTYLUS* Raffles 1821

Reproduced by courtesy of Rühmekorf, Hannover

GEOG DIST: Sumatra, Malay Peninsula   BDY WT: 9500–12 700g♂; 9000–11 600g♀
HD/BDY LTH: 468–595 mm♂; 438–630 mm♀ TAIL LTH:
COAT COL: black

SYNONYMS

VERNACULAR NAMES

Siamang
Great gibbon

3.4.1.1.   *PONGO PYGMAEUS*   (Linnaeus 1760)

Reproduced by courtesy of Bernhard Grzimek

GEOG DIST: Sumatra, Borneo

HD/BDY LTH: mean 965 mm♂; 768 mm♀
COAT COL: reddish

MEAN BDY WT: 100 kg♂; 81 kg♀
(Sumatran Orang: 69kg♂; 37kg♀)
MEAN HT: 1370 mm♂; 1150 mm♀

SYNONYMS

VERNACULAR NAMES

Orang-utan
Orang-outan

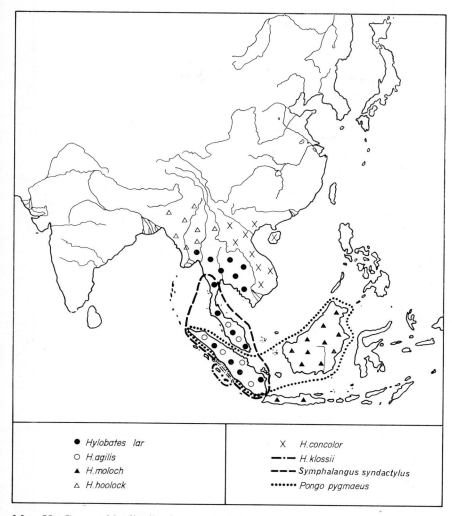

●	Hylobates lar	X	H. concolor
○	H. agilis	—·—	H. klossii
▲	H. moloch	— — —	Symphalangus syndactylus
△	H. hoolock	·······	Pongo pygmaeus

MAP 32. Geographic distribution of *Symphalangus, Pongo* and the different species of *Hylobates* (redrawn and adapted from Fiedler, 1956 and Napier, 1967)

3.4.2.1. *PAN TROGLODYTES* (Blumenbach 1799)

Reproduced by courtesy of Giardino Zoologico, Torino

GEOG DIST: Central Africa
HD/BDY LTH: 1000–1700 mm
COAT COL: black

MEAN BDY WT: 48.9 kg♂; 40.6kg♀
TAIL LTH: no tail

3.4.2.1.　*PAN TROGLODYTES*　(Blumenbach 1799)

SYNONYMS

Troglodytes niger Geoffroy 1812
Pan africanus Oken 1816
Troglodytes leucoprymnus Lesson 1831
Troglodytes tschego Duvernoy 1855
Troglodytes calvus Du Chaillu 1860
Troglodytes vellerosus Gray 1862
Simia ituricus Rassen von Matschie 1912
Anthropopithecus ituricus Rassen von Matschie 1912
Anthropopithecus graueri Matschie 1914
Troglodytes kooloo-kamba Du Chaillu 1860

VERNACULAR NAMES

Chimpanzee
Soko-mutu
Tschego
Scimpanzé
Soko
Kitera
Ekikuya
Ekibandu
Kinyamusito
Esike
Empundu
Mkiba

Reproduced by courtesy of Photo V. Six, Antwerp Zoo

GEOG DIST: Congo
HD/BDY LTH: *ca* 900 mm
COAT COL: Black

BDY WT: 48·9kg♂; 49·6kg♀
TAIL LTH: no tail

3.4.2.2.   *PAN PANISCUS*   Schwarz 1929

VERNACULAR NAMES

Bonobo
Lesser chimpanzee
Chimpanzé nain
Pygmy chimpanzee
Swergchimpanzee
Zwergschimpanse

3.4.3.1. *GORILLA GORILLA* (Savage and Wyman 1847)

Photo Bernhard Grzimek, reproduced by courtesy of Zoological Society of Philadelphia

3.4.3.1. *GORILLA GORILLA* (Savage and Wyman 1847)

GEOG DIST: Equatorial Africa
HD/BDY LTH: 1400–1850 mm
COAT COL: grey black

BDY WT: 140–275kg♂; 70–120kg♀
TAIL LTH: no tail

SYNONYMS

Troglodytes savagei Owen 1848
Gorilla gina Geoffroy 1855
Gorilla castaneiceps Slack 1862
Gorilla mayéma Alix and Bouvier 1877
Gorilla graueri Matschie 1903
Pan gorilla

VERNACULAR NAMES

Flachlandgorilla
Coast gorilla
Gorille du côte
Berggorilla
Mountain gorilla
Gorille de montagne
Makaku
Ngagi

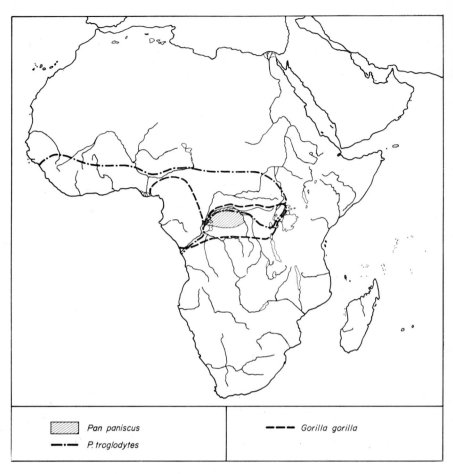

MAP 33. Geographic distribution of *Gorilla* and the different species of *Pan* (redrawn adapted from Fiedler, 1956 and Napier, 1967)

The Man of Leonardo da Vinci

SYNONYMS

The synonyms for man have been selected from the unpublished check-list of A. Simonetta to show that, even for our own species, disagreement existed among different scientists as to its proper taxonomic classification.

Homo sapiens Linnaeus 1788
Homo troglodytes Linnaeus 1788
H. sinicus Bory Saint-Vincent 1825
H. scythieus Bory Saint-Vincent 1825
H. neptunianus Bory Saint-Vincent 1825
H. patagonus Bory Saint-Vincent 1825
H. melaninus Bory Saint-Vincent 1825
H. japeticus Bory Saint-Vincent 1825
H. indicus Bory Saint-Vincent 1825
H. hyperboreus Bory Saint-Vincent 1825
H. hottentotus Bory Saint-Vincent 1825
H. columbicus Bory Saint-Vincent 1825
H. cafer Bory Saint-Vincent 1825
H. australasiensis Bory Saint-Vincent 1825

H. aethiopicus Bory Saint-Vincent 1825
H. arabicus Bory Saint-Vincent 1825
Euplokamus mediderraneus Hoeckel 1898
Euplokamus dravida Hoeckel 1898
Euplokamus nuba Hoeckel 1898
Euplokamus australis Hoeckel 1898
Euthycomus mongolus Hoeckel 1898
Euthycomus malayus Hoeckel 1898
Euthycomus arcticus Hoeckel 1898
Euthycomus australis Hoeckel 1898
Laphocomus papua Hoeckel 1898
Heoanthropus eurasieus G. Sergi 1911
Heoanthropus orientalis G. Sergi 1911
Heoanthropus subarcticus G. Sergi 1911
Heoanthropus arcticus G. Sergi 1911
Hesperanthropus columbi G. Sergi 1911
Hesperanthropus patagonicus G. Sergi 1911
Hesperanthropus oceanicus G. Sergi 1911
Hesperanthropus tasmanianus G. Sergi 1911
Homo hodiernus Schwalbe 1902
Homo euro-africanus Fisher 1830
Homo semiticus Fisher 1830
Homo austro-africanus Fisher 1830
Homo oceanius Fisher 1830
Homo polynesius Fisher 1830
Homo novo-hollandus Fisher 1830
Homo curilanus Fisher 1830
Homo brachycephalus Wilser 1903
Homo perniciosus Girault 1924
Homo caesicus Bryn 1929
Homo flatulens Hooten 1939

INDEX OF SPECIES

Page numbers in italics indicate distribution maps

337

INDEX OF SYNONYMS

A

Allenopithecus nigroviridis, 3.1.5.21.
Alouatta aequatorialis, 2.3.6.1.
Alouatta guariba, 2.3.6.2.
Alouatta insulanus, 2.3.6.3.
Alouatta jaura, 2.3.6.3.
Alouatta macconnelli, 2.3.6.3.
Alouatta nigerrima, 2.3.6.4.
Alouatta nigra, 2.3.6.5.
Alouatta palliata, 2.3.6.1.
Alouatta ursina, 2.3.6.2.
Altililemur medius, 1.4.2.2.
Aluatta ursina, 2.3.6.2.
Ameranthropoides loysi, 2.3.9.3.
Anathana pallida, 1.1.1.9.
Anathana wroughtoni, 1.1.1.9.
Anthropopithecus graueri, 3.4.2.1.
Anthropopithecus ituricus, 3.4.2.1.
Aotes aversus, 2.3.1.1.
Aotes azarae, 2.3.1.1.
Aotes bidentatus, 2.3.1.1.
Aotes bipunctatus, 2.3.1.1.
Aotes boliviensis, 2.3.1.1.
Aotes duruculi, 2.3.1.1.
Aotes griseimembra, 2.3.1.1.
Aotes gularis, 2.3.1.1.
Aotes humboldtii, 2.3.1.1.
Aotes lanius, 2.3.1.1.
Aotes lemurinus, 2.3.1.1.
Aotes miconax, 2.3.1.1.
Aotes microdon, 2.3.1.1.
Aotes miriquina, 2.3.1.1.
Aotes nigriceps, 2.3.1.1.
Aotes oseryi, 2.3.1.1.
Aotes pervigilis, 2.3.1.1.
Aotes roberti, 2.3.1.1.
Aotes rufipes, 2.3.1.1.
Aotes senex, 2.3.1.1.
Aotes spixii, 2.3.1.1.
Aotes vociferans, 2.3.1.1.
Aotes zonalis, 2.3.1.1.
Aotus azarae, 2.3.1.1.
Ateles arachnoides, 2.3.10.1.
Ateles ater, 2.3.9.1.
Ateles bartlettii, 2.3.9.3.
Ateles chuva, 2.3.9.3.
Ateles cucullatus, 2.3.9.4.
Ateles dariensis, 2.3.9.2.
Ateles eriodes, 2.3.10.1.

Ateles fuliginosus, 2.3.9.3.
Ateles grisescens, 2.3.9.4.
Ateles hemidactilus, 2.3.10.1.
Ateles hybridus, 2.3.9.3.
Ateles hypoxanthus, 2.3.10.1.
Ateles longimembris, 2.3.9.1.
Ateles marginatus, 2.3.9.3.
Ateles marimonda, 2.3.9.3.
Ateles melanochir, 2.3.9.4.
Ateles neglectus, 2.3.9.4.
Ateles niger, 2.3.9.1.
Ateles ornatus, 2.3.9.4.
Ateles pan, 2.3.9.4.
Ateles pentadactylus, 2.3.9.1.
Ateles problema, 2.3.9.3.
Ateles robustus, 2.3.9.2.
Ateles rufiventris, 2.3.9.4.
Ateles subpentadactylus, 2.3.9.1.
Ateles tricolor, 2.3.9.4.
Ateles variegatus, 2.3.9.3.
Ateles vellerosus, 2.3.9.4.
Avahis laniger, 1.5.2.1.

B

Brachuteleus arachnoides, 2.3.10.1.
Brachyteles frontatus, 2.3.9.4.
Brachyteles macrotarsus, 2.3.10.1.
Brachyurus calvus, 2.3.5.1.
Brachyurus israelitica, 2.3.4.1.
Brachyurus melanocephalus, 2.3.5.3.
Brachyurus ouakary, 2.3.5.3.
Brachyurus rubicundus, 2.3.5.2.

C

Cacajao roosevelti, 2.3.5.1.
Callicebus brunneus, 2.3.2.1.
Callicebus calligatus, 2.3.2.1.
Callicebus cinerascens, 2.3.2.1.
Callicebus emiliae, 2.3.2.1.
Callicebus hoffmannsi, 2.3.2.1.
Callicebus lucifer, 2.3.2.2.
Callicebus melanops, 2.3.2.1.
Callicebus moloch, 2.3.2.1.
Callicebus ollalae, 2.3.2.1.
Callicebus personatus, 2.3.2.1.
Callicebus remulus, 2.3.2.1.
Callicebus ustofuscus, 2.3.2.1.
Callimico snethlageri, 2.2.1.1.
Callimidas snethlageri, 2.2.1.1.

345

INDEX OF VERNACULAR NAMES